Introductory Muon Science

Muons are unstable elementary particles that are found in space, and which can also be produced with a billion times that intensity by particle accelerators. This book describes the various applications of muons across the spectrum of the sciences and engineering.

Scientific research using muons relies both on the basic properties of the particle and also the microscopic (at the atomic level) interaction between muons and surrounding particles such as nuclei, electrons, atoms, and molecules. Examples of research that can be carried out using muons include muon catalysis for nuclear fusion, the application of muon spin probes to study microscopic magnetic properties of advanced materials, electron labeling to help in the understanding at the microscopic level of electron transfer in proteins, and nondestructive elemental analysis of the human body. Cosmic-ray muons can even be used to study the inner structure of volcanoes.

This fascinating summary of muon science will be of interest to physicists, materials scientists, chemists, biologists, and geophysicists who want to know how what has come to be known as the particle of the twenty-first century can be used in their areas of research.

Introductory Muon Science

Kanetada Nagamine

High Energy Accelerator Research Organization, Tsukuba, Ibaraki, Japan

CAMBRIDGE
UNIVERSITY PRESS

CAMBRIDGE UNIVERSITY PRESS
Cambridge, New York, Melbourne, Madrid, Cape Town, Singapore, São Paulo

Cambridge University Press
The Edinburgh Building, Cambridge CB2 8RU, UK

Published in the United States of America by Cambridge University Press, New York

www.cambridge.org
Information on this title: www.cambridge.org/9780521593793

First published 2003
This digitally printed version 2007

A catalogue record for this publication is available from the British Library

ISBN 978-0-521-59379-3 hardback
ISBN 978-0-521-03820-1 paperback

Contents

Preface

Since the discovery of cosmic rays in 1940, elementary particle muons have become fascinating and exotic particles which can be objectives and/or tools for fundamental physics and applied science. In particular, after intense muons became available by using particle accelerators in 1960, the field of scientific research using muons has been growing year by year.

From the author's viewpoint, three major unique features of muons have formed the basis of all muon-related scientific research: (1) unique mass, such as heavy electrons and light protons; (2) radioactivity with polarization phenomena; and (3) the electromagnetic interaction nature with matter without a strong interaction. These features have promoted the application of muons to: (1) muon catalyzed fusion for future atomic energy; (2) sensitive probes of the microscopic magnetic properties of new materials and biomolecules; and (3) radiography of a large-scale substance for preventing natural disasters, respectively.

The author made special efforts for this book to include a self-contained description of all physics principles required for muon applications. In these applications, we note that muons may be the key particles to provide answers to the basic problems associated with possible crises in human life in the twenty-first century, namely, a shortage of energy resources, the need for more information on the biological functioning of the human body, and the need to prevent natural disasters, such as volcanic eruptions and earthquakes. Complete descriptions are given of ways to apply elementary particle muons to these three major human problems. Instructions are also given to young scientists and/or students about how a basic understanding of fundamental physics can contribute to the growth of human life.

Thus, the academic disciplines of this book can be listed as follows:

1. atomic energy studies, particularly of fusion energy
2. condensed-matter studies of advanced materials
3. life science studies, particularly of the biological functioning of macromolecules
4. geophysical studies related to the prevention of natural disasters, such as volcanic eruptions and earthquakes

The most fundamental properties of muons μ^+, μ^-, and muonium (Mu) are described in Chapter 1 and the sources of muons from both accelerators and cosmic rays are given in Chapter 2, followed by the states of the muon in matter and the thermalization/formation process in Chapter 3. Then, in Chapters 4 and 5, the fundamental properties of muonic atoms and full descriptions of muon catalyzed fusion phenomena and their applications are given, respectively. Then, the subject changes to µSR (muon spin rotational/relaxation/resonance) spectroscopy, describing the principle in Chapter 6 and its probe nature of host materials as an unperturbed probe in Chapter 7, where, because of the wide variety of activities, existing worldwide selected topics are limited to the author's related field. The μSR studies concerning the active-probe nature for creating new microscopic systems follow in Chapter 8, including biological applications. The relatively young subject of cosmic-ray muon radiography is discussed in Chapter 9, followed by future directions from the author's point of view in Chapter 10.

The author would like to express sincere thanks to the following people who contributed to the content of each chapter of this book including supply of beautiful figures before publication:

A. P. Mills, Jr, S. Chu, S. N. Nakamura, Y. Miyake, J. M. Poutissou, and K. Jüngmann (Chapter 1)

K. Ishida, T. Matsuzaki, J. Imazato, H. Nakayama, Y. Miyake, K. Shimomura, K. Nishiyama, A. P. Mills, Jr, G. M. Marshal, E. Morenzoni, and Y. Kuno (Chapter 2)

V. E. Storchak, H. Tanaka, and J. H. Brewer (Chapter 3)

L. I. Ponomarev, T. Koike, S. N. Nakamura, W. Kutschera, K. Sakamoto, M. Oda, and P. Strasser (Chapter 4)

T. Matsuzaki, K. Ishida, N. Kawamura, S. N. Nakamura, M. Kamimura, Y. Kino, S. E. Jones, M. Leon, L. I. Ponomarev, M. Faifman, L. Bogdanova, and J. S. Cohen (Chapter 5)

T. Yamazaki, K. Nishiyama, N. Nishida, E. Torikai, J. H. Brewer, R. H. Kiefl, and A. Schenck (Chapter 6)

R. Kadono, W. Higemoto, N. Nishida, J. H. Brewer, Y. J. Uemura, I. Watanabe, Y. Koike, E. Torikai, A. Ito, and K. Kojima (Chapter 7)

R. Kadono, J. Kondo, K. Ishida, H. Shirakawa, F. L. Pratt, E. Torikai, S. Ohira, I. Watanabe, N. Go, and D. G. Fleming (Chapter 8)

K. Ishida, H. Tanaka, S. N. Nakamura, K. Shimomura, M. Iwasaki, M. Oda, Y. Totsuka, H. Wakita, and Y. Ida (Chapter 9)

P. Bakule, Y. Miyake, K. Shimomura, and Y. Mori (Chapter 10)

Careful proof-reading by Drs J. M. Poutissou (Chapter 1), K. Ishida (Chapters 2 and 3), L. I. Ponomarev (Chapters 4 and 5), A. Schenck and T. P. Das (Chapters 6–8) and P. Bakule (Chapter 10) is acknowledged. Careful reading of the whole content with valuable comments and revision of English by Dr R. Macrae is also acknowledged. This work could only be completed with the help of the wonderful keyboard skills of Mrs H. Shiosaka, J. Ogata, S. Satoh, and J. Nakai, to whom the author would like to express his sincere thanks.

The author's involvement in muon science research began in 1971. A special acknowledgment is given to Professor T. Yamazaki, who introduced me to the field of muon physics with his long-term collaboration and encouragement. The author's research activities have been supported and encouraged by leading members of the institutions with which the author was associated, to whom sincere thanks are given: Professors T. Yamazaki, A. Arima, T. Nishikawa, H. Sugawara, Y. Kimura, the late M. Oda, S. Kobayashi, E. W. Vogt, and P. W. Williams.

Finally, the author thanks his late parents, Tadao and Masako Nagamine, to whom this book is dedicated.

Abbreviations

μCF	muon catalyzed fusion
μSR	muon spin rotation/relaxation/resonance
AF	antiferromagnetic phase
ALC	avoided-level crossing
bcc	body-centered cubic
BCS	Bardeen–Cooper–Schrieffer
BSCCO	$Bi_2Sr_2CaCuO_{8+\delta}$
BNL AGS	Brookhaven National Laboratory, Alternating Gradient Synchrotron
BNL	Brookhaven National Laboratory
CCD	charge-coupled devices
CCS-11	microprogrammed CAMAC processor
CERN	European Organization for Nuclear Research
CERN SPS	European Organization for Nuclear Research Super Proton Synchrotron
CFD	constant fraction discriminator
CPT	charge transformation, parity inversion, and time reversal
DMAC	direct memory access
DSC	discriminator
EM	electromagnetic
EMC	European Muon Collaboration
ESR	electron spin resonance
ESS	European Spallation Neutron Source
fcc	face-centered cubic
FFAG	fixed-field alternating-gradient synchrotron
FNAL-MI	Fermi National Accelerator Laboratory–Main Injector
FTM	Fourier transformation
hcp	hexagonal close-packed
Hot W	hot tungsten
HTCSC	high-T_c superconductors
HTF-μSR	high transverse field–muon spin rotation
HV	high-voltage power supply
IPNS	Intense Pulsed Neutron Source

IR	interrupt register
ISIS-RAL	ISIS–Rutherford Appleton Laboratory
JAERI-KEK	Japan Atomic Energy Research Institute–High Energy Accelerator Research Organization
JINR	Joint Institute for Nuclear Research
J-PARC	Japan Proton Accelerator Research Complex
KEK-PS	High Energy Accelerator Research Organization–12 GeV Proton Synchrotron
KEK	High Energy Accelerator Research Organization
KEK-MSL	High Energy Accelerator Research Organization–Meson Science Laboratory
LAMPF PSR	Los Alamos Meson Physics Facility Proton Storage Ring
LAMPF	Los Alamos Meson Physics Facility
LBL	Lawrence Berkeley Laboratory
LCR	level-crossing resonance
LF-μSR	longitudinal field μSR
LFNC	Lepton Flavor Non-conservation
LHD	liquid hydrogen density unit
LHe	liquid helium
LN$_2$	liquid nitrogen
LSCO	La$_{1-x}$Sr$_x$CuO$_4$
LTO	low-temperature orthorhombic
LTT	low-temperature tetragonal
MCP	multichannel plate
MS TDC	multistop time-to-digital converter
MuSR	muonium spin rotation
NEQ	nuclear electric quadrupole moment
NMR	nuclear magnetic resonance
PIXE	proton-induced X-ray element analysis
PM	photomultiplier
PMT	photomultiplier tube
PRISM	Phase Rotated Intense Slow Muon Source
PSI	Paul Scherrer Institute
QED	quantum electrodynamics
r.f.	radiofrequency
RAL	Rutherford Appleton Laboratory
RIKEN-RAL	Institute of Physical and Chemical Research–Rutherford Appleton Laboratory branch
SEC	Secondary Enclosure Clean-up System
SIMS	secondary ion mass spectrometry
SNS	Spallation Neutron Source
TDC	time-to-digital converter
TF-μSR	transverse field muon spin rotation

TGHS	tritium gas-handling system
TRIUMF	Tri-University Meson Facility
TRIUMF-UBC	Tri-University Meson Facility–University of British Columbia
UHV	ultrahigh vacuum
UT-TRIUMF	University of Tokyo Group at TRIUMF
V-A theory	vector–axial vector interaction theory
VUV	vacuum ultraviolet
YBCO	$YBa_2Cu_3O_{7-\delta}$
ZF-μSR	zero external field-μSR

1

What are muons? What is muon science?

Scientific research using the fundamental particle known as the muon depends upon the muon's basic particle properties and also on the microscopic (atomic-level) interactions of muons with surrounding particles such as nuclei, electrons, atoms, and molecules. This chapter deals mainly with the fundamental properties of muons based on what is presently known from particle physics. Several relevant reference works exist, in particular regarding historical developments (Hughes and Kinoshita, 1977; Kinoshita, 1990).

1.1 Basic properties of the muon

In one sentence, the properties of muons can be summarized as follows:

Muons are unstable elementary particles of two charge types (positive μ^+ and negative μ^-) having a spin of 1/2, an unusual mass intermediate between the proton mass and the electron mass (1/9 m_p, 207 m_e), and 2.2 µs lifetime.

Over time, a deeper understanding of the above statement has been gained through the development of experimental methods and improvements in theoretical models. Some data relevant to muon science are summarized in Table 1.1.

The uniqueness of lifetime and mass can be understood by comparing muon values to those of other particles, as seen in Figure 1.1. These properties can be summarized as follows:

The muon has the second longest lifetime among all the fundamental unstable particles (that is, omitting particles believed to be stable, such as the proton, electron, and neutrino) after the neutron, and has the second smallest mass among all the fundamental particles after the electron.

The following paragraphs elaborate and clarify the contents of Table 1.1.

1.1.1 Mass of the muon

The most accurate determination of the mass of the muon (m_{μ^+}, m_{μ^-}) with reference to the electron mass (m_e), which is known to a precision of 10^{-8} (10 p.p.b.), can be made in the following two ways.

Table 1.1 Fundamental properties of muons

	μ^+	μ^-
Charge	+1	−1
Mass	206.768 277 (24) $(m_e)^a$	105.659 (1) $(\text{MeV}/c^2)^b$
Spin	1/2	1/2
Magnetic moment (μ_μ/μ_p)	3.183 345 13 $(39)^a$	
Gyromagnetic ratio	13.553 42	
$(\mu_\mu/2\pi I$ in kH_z/G, $I = 1/2)$		
Gyromagnetic factor $(g/2)$	1.001 165 920 3 $(15)^c$	1.001 165 936 (12)
$(\mu_\mu = g(e\hbar/2m_\mu)I$, $I = 1/2)$		
Free decay lifetime $(10^{-6}$ s)	2.196 95 $(6)^d$	2.194 8 $(10)^f$
	2.197 078 $(73)^e$	(in flight)
Decay mode	$e^+ + \bar{\nu}_\mu + \nu_e$	(100%)
	$e^+ + \gamma$	$(<1.7 \times 10^{-10})$
	$e^+ + e^- + e^+$	$(<1.9 \times 10^{-9})$
	$e^+ + \gamma + \gamma$	$(<1.25 \times 10^{-8})$

^{a} Liu et al., 1999; ^{b} Beltrami et al., 1986; ^{c} Brown et al., 2001; ^{d} Giovanetti et al., 1984; ^{e} Bardin et al., 1984; ^{f} Williams and Williams, 1972.

As will be described more fully later, the mass of the positive muon ($m_{\mu+}$) can be determined most accurately by measuring the energy interval between the 1s and 2s electronic quantum levels (ΔE_{1s-2s}) in muonium (the neutral bound state of μ^+ and e^-, closely analogous to the hydrogen atom, sometimes designated hereafter as Mu). Laser two-photon resonance spectroscopy gives an isotope shift in the ΔE_{1s-2s} absorption line for Mu with respect to H due to a change in the reduced mass correction.

The mass of the negative muon ($m_{\mu-}$), on the other hand, cannot be determined in this way, but can be obtained by measuring the energy intervals between atomic states formed by the μ^- around a nucleus. Such a system is called a muonic atom. Because the mass of the μ^- is 207 times that of the electron and thus much closer to the nuclear mass than those of the orbiting electrons in a conventional atom, specific atomic states should be carefully selected such that ambiguity related to the nuclear charge distribution (i.e. due to the fact that the nucleus is not point-like) is minimized. The actual determination was made by measuring 3d–2p transitions in muonic ^{28}Si with a careful correction (Beltrami et al., 1986).

1.1.2 Lifetime of the muon

The μ^+ and μ^- in vacuum have the following major (100%) decay modes:

$$\mu^+ \to e^+ + \bar{\nu}_\mu + \nu_e$$
$$\mu^- \to e^- + \nu_\mu + \bar{\nu}_e$$

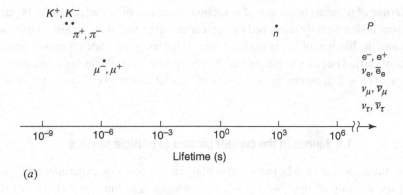

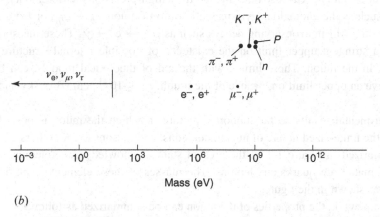

Figure 1.1 Lifetimes of various particles (a) and their masses (b).

where ν_e and ν_μ are the electron and muon neutrinos and $\bar{\nu}_e$ and $\bar{\nu}_\mu$ the corresponding antineutrinos.

The positive muon lifetime, τ_μ, can be measured from the shape of the time spectrum of the decay positrons with reference to the time of μ^+ stopping in some target material under the reasonable assumption that the decay mode of μ^+ in matter is not subject to any changes from that in vacuum. The time distribution of decay positrons $N_e(t)$ follows an exponential law:

$$N_e(t) = N_e(0)\, e^{-t/\tau_\mu}$$

After several attempts at various accelerator facilities worldwide, it was recognized that the removal of spin polarization-related effects is essential. One of the most reliable ways to do this is to measure the decay e^+ over the full 4π solid angle. Several subsequent trials have been conducted (Bardin *et al.*, 1984; Giovanetti *et al.*, 1984; Nakamura *et al.*, 1999).

The lifetime of μ^- must be measured in vacuum since that of bound μ^- in the 1s orbit of a muonic atom is significantly shortened by nuclear capture processes. An alternative method is to measure the lifetime of μ^- in flight compared to that of μ^+. Such a measurement was successfully carried out as a byproduct of the measurement of the muon anomalous magnetic moment $(g - 2)_\mu$, assuming the validity of special relativity (Williams and Williams, 1972).

1.2 Muons in the current picture of particle physics

The size of the μ^+ and μ^- can be measured in high-energy collision experiments using e^+e^- colliders; the reaction $e^+ + e^- \rightarrow \mu^+ + \mu^-$, assuming quantum electrodynamics (QED) and with point-like e^+ and e^-, confirms that μ^+ and μ^-, too, are point-like, with $r_\mu \leq 10^{-16}$cm (Martyn, 1990).

The size of μ^+ and μ^- can also be estimated from high-precision measurements of muon properties such as the anomalous gyromagnetic ratio of muon, $(g - 2)_\mu$, or from the upper limit upon flavor nonconserving decays such as $\mu^+ \rightarrow e^+ + \gamma$. These measurements also place a stringent upper limit on the existence of possible internal structure or excited states in the muon. These limits, with the aid of theoretical models, can be converted to give an upper limit on the size of the muon: $r_\mu \leq 10^{-17}$ cm (Brodsky and Drell, 1980).

All experimental results so far support a picture in which the muon is point-like, in contrast to the finite-sized nature of nuclei, nucleons, and mesons (π, K, others).

As summarized in Figure 1.2, at the present state of knowledge, the elementary constituents of matter are quarks and leptons. The masses of these elementary particles are distributed as shown in the figure.

In this framework, the properties of the muon can be summarized as follows:

The muon belongs to the lepton family, along with the electron, the τ-particle, and their corresponding neutrinos (ν_e, ν_μ, and ν_τ). The muon interacts with other particles and matter through both electromagnetic and weak interactions.

The classification of these elementary particles, as shown in Figure 1.2, is strictly determined by the conservation laws; generations cannot be mixed. Thus, reactions such as $\mu^+ \rightarrow e^+ + \gamma$ or $\mu^- + A \rightarrow e^- + A$ are forbidden. As described later, in the search for violation of generation conservation, the conservation or nonconservation of flavor is one of the central subjects in current particle physics.

1.3 Fundamental interactions of the muon

The μ^+ and μ^- are subject to electromagnetic and weak interactions. These two interactions are now unified into an electroweak interaction within the framework of the standard model. In the following section, these fundamental interactions, which appear again and again in muon science studies, are summarized.

		Generation			Charges			Interactions				
		1	2	3	Electric	Weak	Color	Electric	Weak	Strong	"Mass"	Gravity
"Matter" particles	Quarks	Up	Charm	Top	2/3	↑	Red	◯	◯	◯	◯	◯
		Up	Charm	Top			Green					
		Up	Charm	Top			Blue					
		Up-type quarks										
		Down	Strange	Bottom	−1/3	↓	Red	◯	◯	◯	◯	◯
		Down	Strange	Bottom			Green					
		Down	Strange	Bottom			Blue					
		Down-type quarks										
	Leptons	ν_e	ν_μ	ν_τ	0	↑	White	✕	◯	✕	✕ ?	◯ ?
		Neutrinos										
		Electrons	μ	τ	−1	↓	White	◯	◯	✕	◯	◯ ?
		Charged leptons										
"Force" particles								γ (photon)	W, Z	Gluons	Higgs ?	Gravitons ?
									Gauge bosons			

(a)

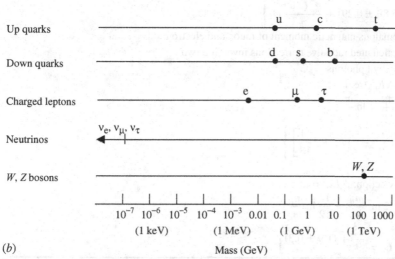

(b) Mass (GeV)

Figure 1.2 Basic properties of quarks, leptons, and gauge particles (a), and their masses (b).

1.3.1 Electromagnetic (EM) interaction

Both charge types of the muon interact with other charged particles via the Coulomb interaction in which the potential energy is given by $-e^2 Z/r$, where Z is the charge of the other particle (the charge on the muon being ± 1). Several important atomic bound states are formed, including: muonium ($\mu^+ e^-$), muonic hydrogen ($\mu^- p$), and muonic Z-atoms ($\mu^- Z$).

The magnetic moments of μ^+ and μ^- (μ_μ) interact with magnetic fields either intrinsic to the atoms themselves or externally applied. The hyperfine splittings in the atomic bound states and the spin precession frequencies around the external field (H_{ext}) are thus determined by the relevant parameters:

$$\Delta E_{\text{hfs}}(\text{Mu}, 1\text{s}) = \mu_\mu \times \mu_e <1/r^3>, \qquad f_\mu = \gamma_{\text{Mu}}(= \mu_\mu/2\pi)H_{\text{ext}}$$

Table 1.2 Hyperfine splitting of the muonium ground state $\Delta\nu = \Delta E_{\text{hfs}}(\text{Mu, 1s})/h$; terms of the theoretical predictions and experiment (Sapirstein and Yennie, 1990)

Theory	Experiment
4 463 303.11(1.33)(0.40)(1.0) kHz	4 463 302.765(53) kHz
Fermi splitting; E_F	(Liu *et al.*, 1999)

$$\frac{16}{3}\alpha^2 \frac{m_r^3}{m_e^2 m_\mu} hc R_\infty$$

where $m_r = m_e/(1 + m_e/m_\mu)$

R_∞ = Rydberg constant

$\frac{8}{3} \times [1\,233\,690\,735.4(1)\text{ MHz}]$

QED correction: $\Delta E_{\text{hfs}}(\text{QED})$

$$E_F\left(1 + a_\mu\right)\left\{1 + \frac{3}{2}\left(Z\alpha\right)^2 + a_e + \alpha\left(Z\alpha\right)\left(\ln 2 - \frac{5}{2}\right)\right.$$

$$-\frac{8\alpha\left(Z\alpha\right)^2}{3\pi}\ln Z\alpha\left(\ln Z\alpha - \ln 4 + \frac{281}{480}\right)$$

$$\left.+\frac{\alpha\left(Z\alpha\right)^2}{\pi}\left(15.88 \pm 0.29\right) + \frac{\alpha^2\left(Z\alpha\right)}{\pi}D_1\right\}$$

where a_μ, a_e = anomalous magnetic moment of muon and electron

D_1 = uncalculated radiative corrections involving two
virtual photons

Recoil correction: $\Delta E_{\text{hfs}}(\text{rec})$

$$E_F\left\{-\frac{3\alpha}{\pi}\frac{m_e m_\mu}{m_\mu^2 - m_e^2}\ln\frac{m_\mu}{m_e}\right.$$

$$\left.+\frac{\gamma^2}{m_e m_\mu}\left[2\ln\frac{m_r}{2\gamma} - 6\ln 2 + 3\frac{11}{18}\right]\right\}$$

where $\gamma = m_r \alpha$

Radiative-recoil correction: ΔE_{hfs} (rad-rec)

$$E_F\left(\frac{\alpha}{\pi}\right)^2\frac{m_e}{m_\mu}\left[-2\ln^2\frac{m_\mu}{m_e} + \frac{13}{12}\ln\frac{m_\mu}{m_e}\right.$$

$$\left.+\frac{21}{2}\zeta(3) + \frac{\pi^2}{6} + \frac{35}{9} + (1.9 \pm 0.3)\right]$$

QED, quantum electrodynamics.

The energy levels of the bound states of Mu, a centrosymmetric two-particle system very similar to H, have been the objects of several precise measurements. The energy levels of the excited states with reference to the ground state are seen in Figure 1.3. In contrast to H, the core in Mu is truly a single structureless particle, and so the fundamental EM interaction in the two-body bound state can be studied without corrections for core structure; thus, at least in principle, the experimental values of the fundamental parameters of the EM interaction can be obtained more straightforwardly through studies of Mu. The present status of experiment and theory on the Mu ground-state hyperfine energy splitting of the Mu in vacuum is summarized in Table 1.2. The result is presented in terms of $\Delta\nu$ ($\equiv \Delta E_{\text{hfs}}(\text{Mu}/1\text{s})/h$), while, in some other cases, the following expression is used: $\Delta E_{\text{hfs}} = \hbar\omega_0$. The ground state of Mu is subject to energy splitting in an applied external

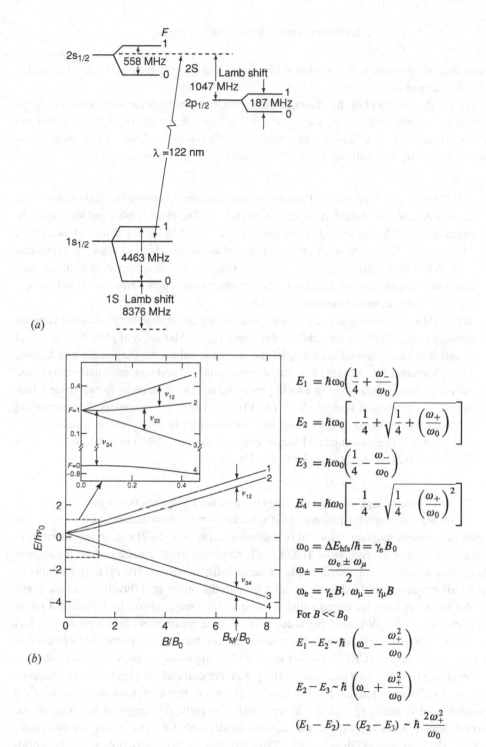

Figure 1.3 Energy levels of Mu, where the lifetime of each state is $\tau_{Mu}(2p) = 1.6$ ns and $\tau_{Mu}(2s) = 0.145$ s (a) and energy diagram of the ground state of Mu against applied external field, so called Breit–Rabi diagram, where energy is in units of $h\nu_0$ ($\nu_0 = 4.463.302$ MHz) and magnetic field is in units of B_0 (0.1585 T) (b).

field; this is expressed in the Breit–Rabi formula. The energy levels and related formulae are summarized in Figure 1.3.

Tests of the validity of the fundamental theory of QED have been carried out through high-precision spectroscopy making use of μ^+- and μ^--containing atoms. These experiments also yield values for the fundamental constants of the muon itself, such as the mass of the muon m_μ and the magnetic moment of the muon μ_μ. To give some examples:

1. ΔE (Mu, 1s–2s): Two-photon laser resonance was carried out for the first time at High Energy Accelerator Research Organization (KEK) (Chu *et al.*, 1988), and subsequently extended at ISIS–Rutherford Appleton Laboratory (ISIS–RAL) (Maas *et al.*, 1996; Meyer *et al.*, 2000). The most updated value of the muon–electron mass ratio obtained from ΔE (Mu, 1s–2s) is $m_{\mu^+}/m_{e^-} = 206.768\,38(17)$, which is consistent with the most accurate value to be mentioned later. This measurement is now known to provide potentially the most accurate determination of m_μ.
2. ΔE_{hfs} (Mu, 1s): Microwave resonance spectroscopy under high magnetic field simultaneously yielded ΔE_{hfs} to provide m_{μ^+}/m_{e^-} and μ_μ/μ_p (Mariam *et al.*, 1982). Theoretical progress towards the understanding of the experimental results (Sapirstein and Yennie, 1990) is summarized in Table 1.2. This experimental method requires a narrowing of the measured line in order to obtain an improved value. There, in order to overcome a limitation due to natural line width of 145 kHz ($\cong 1/\pi\tau_\mu$), a resonance line-narrowing technique is employed by interacting microwave with Mu atoms which have lived several times τ_μ. The most updated measurement (Liu *et al.*, 1999) provided $m_{\mu^+}/m_{e^-} = 206.768\,277(24)$ and $\mu_\mu/\mu_p = 3.183\,345\,13(39)$.

The other type of high-precision measurement of the EM interaction is the measurement of the anomalous magnetic moment of the muon by storing muon motion in a high magnetic field. Anomalous magnetic moment of the muon $a_\mu(= (g-2)/2)$ can be measured from the so-called $(g-2)$ precession. The $(g-2)$ precession which corresponds to the angular frequency difference between the spin precession frequency and the cyclotron frequency in a uniform magnetic field perpendicular to both the muon spin direction and the plane of the orbit has been measured using a muon storage ring, where, by selecting a muon momentum of 1.5 GeV/c, any effects due to the electric confinement field were removed. A precision of 10 p.p.m. was obtained in experiments conducted at the European Organization for Nuclear Research (CERN) (Bailey *et al.*, 1975). Improved measurements with the aim of obtaining $(g-2)_\mu$ to a precision of 0.5 p.p.m. are currently in progress at Brookhaven National Laboratory (BNL). For this level of precision, however, improvements over the present level in both m_μ and μ_μ are required. Currently, the weighted mean of all the experimental results agrees with the standard model with 3.6 ± 4.0 p.p.m. (Brown *et al.*, 2000). The latest report (Brown *et al.*, 2001) provides the comparison between the world-average experimental data and theoretical prediction based on the standard model $a_\mu(\text{exp}) - a_\mu(\text{theory}) = 43(16) \times 10^{-10}$, suggesting an existence of physics beyond the standard model. The more updated report is available (Bennett *et al.*, 2002).

1.3.2 Weak interaction

The weak interaction of the muon is the phenomenon underlying both the decay of μ^+ and μ^- and the nuclear capture of μ^- in muonic atoms. The fundamental law of flavor conservation has been confirmed through observations setting an upper limit on flavor conservation-violating processes such as $\mu^+ \to e^+ + \gamma$ or $\mu^- Z \to e^- Z$.

In addition to lepton number conservation, another important weak-interaction experiment involving muon, muonium, and muonic atom is to search for a conversion of muonium (Mu, $\mu^+ e^-$) to antimuonium ($\overline{\mathrm{Mu}}$, $\mu^- e^+$). This is related to the mixing of lepton numbers, including multiplicative or additive schemes; the standard model in particle physics assumes an additive scheme. Various types of experiments have been done after establishing the experimental method of thermal Mu production in a vacuum. The present experiment gives an upper limit of the conversion probability of $P_{\mathrm{Mu}\overline{\mathrm{Mu}}} \leq 8.3 \times 10^{-11}$ (90% CL) (Willmann et al., 1999).

At the same time, the detailed properties of the normal decay process of the μ^+ yielding an e^+ and two neutrinos have been studied to a high degree of precision. For a purely weak process, muon decay can be written using the four Michel parameters, ρ, η, ε, and δ, as summarized in Figure 1.4 (Michel, 1949; Kinoshita and Sirlin, 1957), where in some cases nonzero parameter values correspond to violation of a fundamental conservation law, again summarized in Figure 1.4.

1.4 Production and decay of polarized muons

Given the present upper limit for flavor nonconserving rare decay processes and the high-precision determination of the normal decay process, the μ^+ production and decay processes can be characterized as follows.

1.4.1 Muon polarization in $\pi\mu$ decay

The muon is produced in the decay of the pion according to:

$$\pi^+ \to \mu^+ + \nu_\mu$$
$$\pi^- \to \mu^- + \overline{\nu}_\mu$$

Since the spin of the pion is zero and the muon neutrino has a definite helicity (h) such that $h = -1$ for $\overline{\nu}_\mu$ and $h = +1$ for ν_μ, the muon is 100% polarized in the center-of-mass system, as shown in Figure 1.5.

1.4.2 Asymmetry of electron/positron emission in muon decay

The muon decays into an electron and two neutrinos as follows:

$$\mu^+ \to e^+ + \nu_e + \overline{\nu}_\mu$$
$$\mu^- \to e^- + \overline{\nu}_e + \nu_\mu$$

$$dN(x,\ \theta) = \frac{d^3p}{(2\pi)^4}\frac{m_\mu E_0}{12}A\left\{6(1-x)+4\rho\left[\frac{4}{3}x-1-\frac{1}{3}\frac{m_e^2}{E_0^2 x}\right]\right.$$

$$+6\eta\frac{m_e}{E_0}\frac{(1-x)}{x}\pm\beta\xi\cos\theta\left[2(1-x)\right.$$

$$\left.\left.+4\delta\left(\frac{4}{3}x-1-\frac{1}{3}\frac{m_e^2}{m_\mu E_0}\right)\right]\right\}$$

upper sign: μ^+ decay, lower sign: μ^- decay

$$A = a + 4b + 6c$$

$$\rho = \frac{3b+6c}{a+4b+6c}$$

$$\eta = \frac{\alpha-2\beta}{a+4b+6c}$$

$$\xi = \frac{-3a'-4b'+14c'}{a+4b+6c}$$

$$\delta = \frac{3b'-6c'}{3a'+4b'-14c'}$$

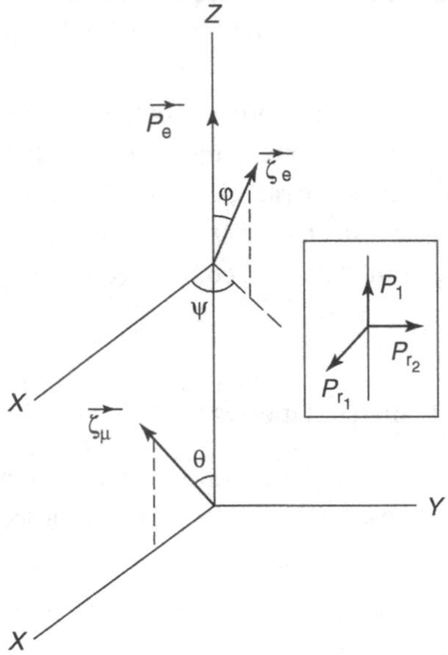

where

$$a = |g_S|^2 + |g_{S'}|^2 + |g_P|^2 + |g_{P'}|^2$$
$$b = |g_V|^2 + |g_{V'}|^2 + |g_A|^2 + |g_{A'}|^2$$
$$c = |g_T|^2 + |g_{T'}|^2$$
$$\alpha = |g_S|^2 + |g_{S'}|^2 - |g_P|^2 - |g_{P'}|^2$$
$$\beta = |g_V|^2 + |g_{V'}|^2 - |g_A|^2 - |g_{A'}|^2$$
$$a' = 2\mathrm{Re}(g_S g_{P'}^* + g_P g_{S'}^*)$$
$$b' = 2\mathrm{Re}(g_V g_{A'}^* + g_A g_{V'}^*)$$
$$c' = 2\mathrm{Re}(g_T g_{T'}^*)$$

$$x = E/E_0$$
$$E_0 = (m_\mu^2 + m_e^2)/2m_\mu$$
$$\cong m_\mu/2$$

Experimental data on decay parameters

$\rho = -0.7518\ (26),$ $\eta = -0.007\ (13)$
$\xi P_\mu \delta/\rho > 0.99682,$ $\delta = 0.779\ (4)$

Figure 1.4 The details of the general formula of normal muon decay where all the possible terms in four Fermion interaction are considered with interaction constants of g_S (scalar), g_V (vector), g_T (tensor), g_A (axial vector), and g_P (pseudoscalar) for the Hamiltonian of $\Sigma_i[(\psi_e \Gamma_i \psi_\mu)(\psi_v(g_i + g_i'\gamma_5)\psi_v) + \text{h.c.}]$ and θ is the angle between electron momentum and muon spin. Experimental data on each parameter are also shown; data from Hagiwara, K. *et al.* (2002). *Phys. Rev.*, **D66**, 010001.

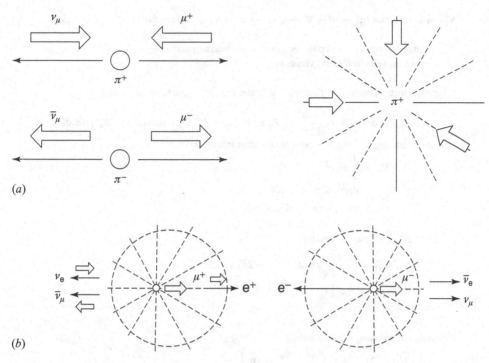

Figure 1.5 (a) Muon spin polarization originating from pion decay. (b) Spatial distribution of positrons from the decay of a polarized muon.

Again, as a result of parity violation, the decay electrons are distributed asymmetrically with respect to muon polarization, as also shown in Figure 1.5. The angular distribution, with respect to the vectors of muon spin $\vec{\sigma}_\mu$ and electron momentum $\vec{p}_e$, is given by:

$$W(\theta) = 1 \mp A \cos\theta_e, \qquad \cos\theta_e(\vec{\sigma}_\mu \times \vec{p}_e)$$
$$\text{upper sign: } \mu^+\text{decay}, \qquad \text{lower sign: } \mu^-\text{decay}$$

As shown in more detail in Figure 1.6, the asymmetry coefficient A depends on the energy of the decay positron from μ^+ decay and the decay electron from μ^- decay. The electron spectrum and the asymmetry parameter are given by the formula:

$$N_e(E_e) \propto [1 \mp A(E_e) \cos\theta_e]$$
$$A(E_e) = (E_e^{\text{max}} - 2E_e)/(3E_e^{\text{max}} - 2E_e)$$
$$\text{upper sign: } \mu^+\text{decay}, \qquad \text{lower sign: } \mu^-\text{decay}$$

where $E_e^{\text{max}} = m_\mu c^2/2 = 53$ MeV. At the endpoint, $E_e = E_e^{\text{max}}$, the decay asymmetry takes its maximum absolute value and is equal to -1.

This can be seen intuitively: at the endpoint (maximum electron energy), the two neutrinos are emitted antiparallel to the electron so that their spins are canceled. Thus, the muon spin is entirely transferred to the electron. Within the standard weak interaction theory, the so-called V–A theory (Michel, 1949; Kinoshita and Sirlin, 1959), the electron also carries

Muon to electron ($\mu^+ \rightarrow e^+ + \bar{\nu}_\mu + \nu_e, \mu^- \rightarrow e^- + \nu_\mu + \bar{\nu}_e$) free decay

Assume V(vector) – A(axial–vector) in weak interaction
Neglect m_e (electron mass) versus m_μ

Positron, electron number in E_e to $E_e + dE_e$ and $d\Omega_e$ at θ_e ($\cos\theta_e = (\vec{p}_e \times \vec{\sigma}_\mu)$)

$$N(E_e, \Omega_e)dE_ed\Omega_e = \frac{g_\mu^2}{12\pi^4\hbar^7c^6} E_m E_e^2 \left[(3E_m - 2E_e) \mp \cos\theta_e (E_m - 2E_e)\right] dE_e d\Omega_e$$

where g_μ = coupling constant of vector interaction

$$E_m = \frac{1}{2} m_\mu c^2$$

upper sign: μ^+ decay

lower sign: μ^- decay

Positron, electron energy spectrum

$$N(E_e)dE_e = \frac{g_\mu^2}{3\pi^3\hbar^7c^6} E_m E_e^2 (3E_m - 2E_e)dE_e$$

$$E_m = \frac{1}{2} m_\mu c^2$$

Positron, electron angular distribution

$$E_e^2 (3E_m - 2E_e) \left[1 \mp \frac{E_m - 2E_e}{3E_m - 2E_e} \cos\theta_e\right]$$

$$[1 \pm \cos\theta_e] \quad \text{for } E_e = E_m$$

$$\left[1 \mp \frac{1}{3} \cos\theta_e\right] \text{for } E_e = 0$$

$$\left[1 \pm \frac{1}{3} \cos\theta_e\right] \text{for positron, election energy average}$$

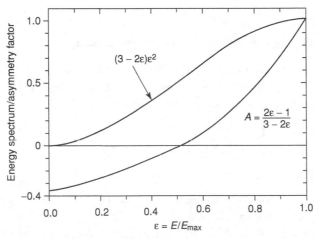

Figure 1.6 An essential part of the formula of muon-to-electron decay according to (vector) – (axial vector) interaction theory.

left-handed helicity. At this energy the electron velocity is so close to the velocity of light that the longitudinal polarization is always equal to the helicity. This means that the electron is preferentially emitted with respect to the muon spin. Usually, in most muon spin rotation/relaxation/resonance (μSR) experiments (described in Chapters 6–8), the electron energy is not selected. In such a case the energy-averaged asymmetry becomes $A = -1/3$, resulting in $W(\theta) = 1 + \frac{1}{3}\cos\theta_e$ for μ^+ decay and $W(\theta) = 1 - \frac{1}{3}\cos\theta_e$ for μ^- decay.

As shown in Chapter 2, spin direction of muon beam depends upon how the beam is produced. Let us summarize here the most popular cases of the stopped muon experiments. For backward decay $\mu^+(\mu^-)$, $\vec{\sigma}_\mu$ of the muon beam is parallel (antiparallel) to the muon momentum vector ($\vec{p}_\mu$). Thus, with respect to muon beam direction, positron (electron) angular distribution is $W(\theta) = 1 + (1/3)\cos\theta_\mu$ ($W(\theta) = 1 + (1/3)\cos\theta_\mu$) (*backward* μ^+, μ^-). For 4 MeV surface μ^+, which is μ^+ from the stopped π^+, just like Figure 1.5, $\vec{\sigma}_\mu$ is opposite to $\vec{p}_\mu$. Thus, with respect to positron angular distribution, $W(\theta) = 1 - (1/3)\cos\theta_\mu$ (*surface* μ^+).

1.5 Other fundamental muon physics

The interaction between the muon and other particles, such as nuclei, nucleons, or quarks, stems mainly from EM and weak interactions, and hence the EM properties of strongly interacting multiquark systems can be probed by very-high-energy ($>$ GeV) muons. The experiments of the so-called European Muon Collaboration (EMC) at CERN, exploring the quark structure of the nucleus, have provided a stimulus for detailed experiments of this type (Berger and Coester, 1987).

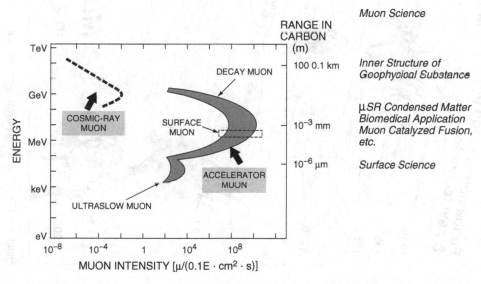

Figure 1.7 Two types of muons (cosmic-ray and accelerator producing) characterized by the ranges of energies, intensities at presently available accelerator facilities stopping range in carbon and related scientific research fields such as μSR, muon spin rotation/relaxation/resonance.

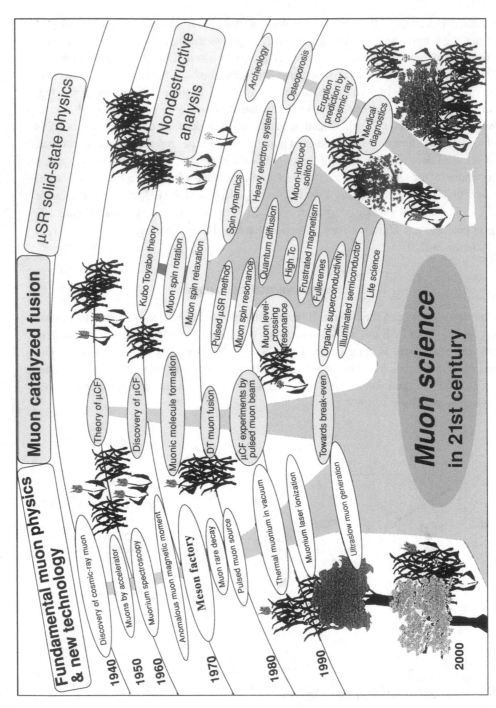

Figure 1.8 Historical development of muon science towards the twenty-first century. μCF, muon catalyzed fusion; μSR, muon spin rotation/relaxation/resonance.

1.6 Muons and muon sciences

Before entering into the subjects related to muon beam production and applications to scientific research, let us overview the existing sources of various types of muons and their applications. As summarized in Figure 1.7, so far, two types of muon beams are available, namely, accelerator-producing muons and cosmic-ray muons. Accelerator-producing muons are high-intensity and low-energy with a short stopping range, while cosmic-ray muons are low-intensity and high-energy with a very long stopping range.

A variety of interesting scientific research has been initiated focusing on these two types of muon beams. The accelerator-producing muons, after stopping in mm–cm-thick target material, are used to conduct condensed-matter studies by the muon spin rotation/relaxation/resonance method, muon catalyzed fusion studies, and nondestructive elemental analysis studies for, e.g., biomedical applications. As a result of development of the ultraslow positive muon technique, sub-μm-thick material can now be an objective of μSR studies. On the other hand, the cosmic-ray muon is now known to be used to measure the density and length of gigantic geophysical substances such as a volcano to learn its inner structure. There is a clear difference between scientific research with accelerator-producing muons and that with cosmic-ray muons; the former mostly concerns experimental studies after stopping the muons inside the objective substance, while the latter is about experimental studies by penetration or scattering.

After the discovery of particles with a mass more than 100 times heavier than that of the electron in the cosmic ray (Anderson and Neddermeyer, 1937, 1938), followed by the discovery of two types of heavy particles, pions and muons (Conversi et al., 1947), the historical development of muon science began. As summarized in Figure 1.8, during two-thirds of the twentieth century, muon science has made remarkable progress in four directions: (1) fundamental muon physics and new technology; (2) muon catalyzed fusion; (3) μSR solid-state physics; and (4) nondestructive analysis. Details of each of these subjects will be presented in the following chapters.

REFERENCES

Anderson, C. D. and Neddermeyer, S. H. (1937). *Phys. Rev.*, **51**, 884.
Anderson, C. D. and Neddermeyer, S. H. (1938). *Phys. Rev.*, **54**, 88.
Bailey, J. *et al.* (1975). *Phys. Lett.*, **B55**, 420.
Bardin, G. *et al.* (1984). *Phys. Lett.*, **137B**, 135.
Beltrami, I. *et al.* (1986). *Nucl. Phys.*, **A451**, 679.
Bennett, G. W. *et al.* (2002). *Phys. Rev. Lett.* **89**, 101804-1
Berger, E. L. and Coester, F. (1987). *Ann. Rev. Nucl. Part. Sci.*, **37**, 463.
Brown, H. N. *et al.* (2000). *Phys. Rev.*, **D62**, 091101.
Brown, H. N. *et al.* (2001). *Phys. Rev. Lett.*, **hep-ex**/0102017.
Brodsky, S. J. and Drell, S. D. (1980). *Phys. Rev.*, **22D**, 2236.
Chu, S. *et al.* (1988). *Phys. Rev. Lett.*, **60**, 101.
Conversi, M. *et al.* (1947). *Phys. Rev.*, **71**, 209.
Giovanetti, K. L. *et al.* (1984). *Phys. Rev.*, **D29**, 343.

Hughes, V. W. and Kinoshita, T. (1977). In *Muon Physics I*, ed. V. W. Hughes and C. S. Wu, p. 11. New York: Academic Press.

Kinoshita, T. (1990). *Quantum Electrodynamics*. Singapore: World Scientific.

Kinoshita, T. and Sirlin, A. (1957). *Phys. Rev.*, **107**, 593; **108**, 844.

Kinoshita, T. and Sirlin, A. (1959). *Phys. Rev.*, **113**, 1652.

Liu, W. *et al.* (1999). *Phys. Rev. Lett.*, **82**, 711.

Maas, F. E. *et al.* (1996). *Phys. Lett.*, **A187**, 247.

Mariam, F. G. *et al.* (1982). *Phys Rev. Lett.*, **49**, 993.

Martyn, H. U. (1990). In *Quantum Electrodynamics* ed. T. Kinoshita, p. 92. Singapore: World Scientific.

Meyer, V. *et al.* (2000). *Phys. Rev. Lett.*, **84**, 49.

Michel, L. (1949). *Proc. Phys. Soc. (Lond.)*, **A63**, 514.

Nakamura, S. N. *et al.* (1999). *RIKEN Rev.*, **20**, 58.

Sapirstein, J. R. and Yennie, D. R. (1990). In *Quantum Electrodynamics*, ed. T. Kinoshita, p. 560. Singapore: World Scientific.

Williams, R. W. and Williams, D. L. (1972). *Phys. Rev.*, **D6**, 737.

Willmann, L. *et al.* (1999). *Phys. Rev. Lett.*, **82**, 49.

2

Muon sources

Experimental studies in muon science can only be conducted when a reasonably intense muon beam of high quality is available. As described in Chapter 1, muons from various sources, including both accelerator-produced particles and those of cosmic-ray origin, are compared in terms of energy and intensity in Figure 1.7. It is easily seen that at very high energy (higher than 100 GeV), cosmic-ray muons are the only possibility, whereas at low energy accelerator-producing muons are almost exclusively used.

In this chapter, more detailed information is given on the types of muon that can be obtained using present production strategies. It should be understood, however, that future developments in both accelerator technology and in ideas for muon beam production are quite likely to change the situation drastically.

2.1 MeV accelerator muons

Muons can only be obtained through the decay of the pions which are produced in nuclear interactions between accelerated particles and nuclear targets. These days, the high-intensity proton accelerator is the most popular source of accelerated particles. Figure 2.1 gives a list of medium-energy proton accelerators currently in use for muon physics research. These days, most activities are based on the accelerators of the so-called meson factories such as the Tri-University Meson Facility (TRIUMF) and Paul Scherrer Institute (PSI) which have beams with an intensity of some 100 µA to mA. From the viewpoint of time structure of accelerators there are two types: continuous and pulsed. At PSI (590 MeV, 1.5 mA) and TRIUMF (500 MeV, 200 µA), the proton beam from a sector focused cyclotron has a continuous character in a macroscopic sense and a microscopic beam structure (intensity modulation) with characteristic frequency 51 MHz at PSI and 23 MHz at TRIUMF. This frequency is simply the radiofrequency (r.f.) of the cyclotron producing the primary beam.

At the High Energy Accelerator Research Organization (KEK), the 500 MeV booster synchrotron for the 12 GeV main ring provides a singly bunched proton beam with 50 ns width and 20 Hz repetition frequency. Its pulse structure is quite unique: the pulse width is much shorter than the muon lifetime τ_μ, while the pulse separation is much longer than τ_μ. This feature provides sharply pulsed muons (Nagamine, 1981). At the ISIS facility of the Rutherford Appleton Laboratory a more advanced synchrotron which generates protons of 800 MeV and 200 µA with a pulse time structure of a double 70 ns (340 ns separation) and

	Energy (GeV)	Particles (per pulse)	Repetition (Hz)	Current (μA)
KEK PS booster	0.5	0.2×10^{13}	20	6
LAMPF PSR	0.8			100
Rutherford ISIS	0.8	2.5×10^{13}	50	200
SNS[a]	1.25			4000
J-PARC-3GeV[a]	3	8.3×10^{13}	25	333
CERN PS	26	0.2×10^{13}	0.50	1.60
KEK PS	12	0.4×10^{13}	0.25	1.16
BNL AGS	30	6×10^{13}	0.30	3.00
Serpukov	70	1.7×10^{13}	0.10	0.27
J-PARC-50GeV[a]	50	30×10^{13}	0.30	15.00

[a]under construction.

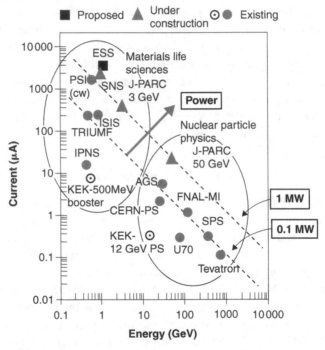

Figure 2.1 Worldwide accelerators of proton synchrotrons capable of intense muon production (above) and current and energy mapping of worldwide proton accelerators (below). ESS, European Spallation Neutrino Source; FNAL-MI, Fermi National Accelerator Laboratory–Main Injector; IPNS, Intense Pulsed Neutron Source; J-PARC, Japan-Proton Accelerator Research Complex; KEK-12 GeV PS, High Energy Accelerator Research Organization–12 GeV Proton Synchrotron; LAMPF, Los Alamos Meson Physics Facility Proton Storage Ring; SNS, Spallation Neutron Source.

50 Hz repetition, originally built for the spallation neutron source, is now in use by a group from the EC for the production of intense pulsed surface μ^+ (Eaton *et al.*, 1988) and by the Institute of Physical and Chemical Research (RIKEN) group (Nagamine *et al.*, 1996; Matsuzaki *et al.*, 2001) for the production of pulsed decay μ^+/μ^- and surface μ^+.

At the beginning of the twenty-first century, there are some demands for the realization of further high-intensity proton accelerators, including: (1) an intense spallation neutron source at the level of 1 MW (e.g., 1 GeV × 1 mA) ~ 10 MW (e.g., 3 GeV × 3.3 mA) for neutron-scattering experiments, radioactive waste disposal, accelerator-driven subcritical reactors, and other applications; and (2) an intense muon source at the MW level to be applied in fields such as $\mu^+\mu^-$ colliders, the search for rare muon decays, a muon-catalyzed fusion-based 14 MeV neutron source, etc.

The main characteristics of these various proton beams which are available now and which are likely to appear in the near future are shown in Figure 2.1, where major present and future proton synchrotron accelerators are presented in terms of proton energy and current. Here, it should be noted that, in addition to the pulsed beam from the synchrotron, there is another type of accelerator known as a fixed-frequency alternating-gradient synchrotron (FFAG) with a higher repetition rate (~ kHz) providing a more favorable pulsed beam for muon science.

As shown schematically in Figure 2.2, the energy of accelerated protons in proton–proton reactions should be greater than twice the pion mass (282 MeV/c^2). Using a nuclear target, this condition becomes somewhat relaxed due to the motional energy of nucleons inside nuclei. Typical examples of pion production cross-section for proton–nucleus reactions are shown in Figure 2.3. A qualitative understanding of the pion production mechanism can be obtained from the Δ-resonance model (Lindenbaum and Sternheimer, 1957), considering that the formation of the Δ-resonance state in nucleon–nucleon scattering is the dominant contribution to production with relevant corrections due to relativistic kinematics and phase-space limitations.

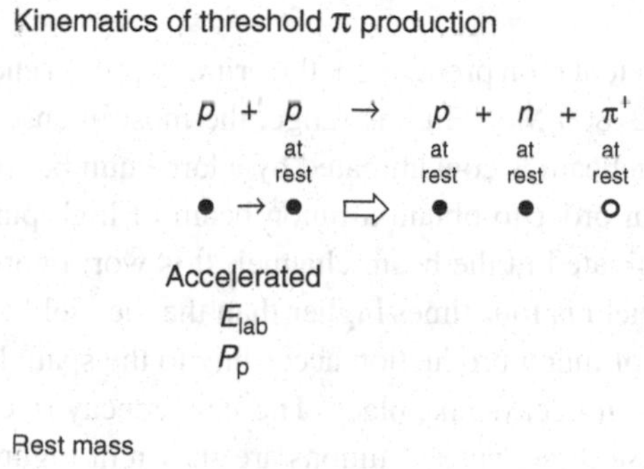

Kinematics of threshold π production

$$p \; + \; p \;\; \rightarrow \;\; p \; + \; n \; + \; \pi^{'}$$

at　　　　　　at　　at　　at
rest　　　　　rest　rest　rest

● → ● ⇨ ● ● ○

Accelerated
E_{lab}
P_p

Rest mass

$$\sqrt{(\varepsilon_p + m_p)^2 - P_p^2} = m_p + m_n + m_\pi$$

$$\sqrt{2m_p^2 + 2m_p\varepsilon_p} = m_p + m_n + m_\pi$$

$$m_p \simeq m_n, \; \varepsilon_p = E_{lab} + m_p$$

$$E_{lab} = \frac{m_\pi^2}{2m_p} + 2m_\pi$$

$$\simeq 2m_\pi$$

Figure 2.2 Kinematic considerations for the required energy of accelerated particles for pion production.

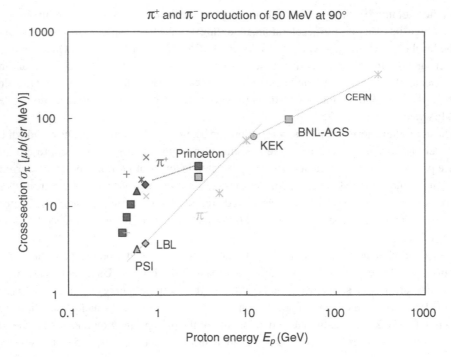

Figure 2.3 Typical cross-sections of pion production in proton–nucleus (W for CERN and carbon for the rest) reactions. PSI, Paul Scherrer Institute; LBL, Lawrence Berkeley Laboratory; KEK, High Energy Accelerator Research Organization; BNL-AGS, Brookhaven National Laboratory Alternating Gradient Synchrotron; CERN, European Organization for Nuclear Research.

For the purpose of efficient pion production, the primary proton energy has typically been chosen to be around 500–800 MeV. In this range, the most intense pions are produced at forward angles. The pion beam is contaminated by a large number of electrons originating from the decay of π^o. In order to obtain a muon beam of high purity, this electron contamination must be eliminated in the beam channel. It is worth noting that, at energies of 500–800 MeV, the π^+ yield is four times higher than the the yield of π^-.

There are three types of muon production according to the spatial position (with respect to the target) where $\pi \rightarrow \mu$ decays take place. These are: decay μ, cloud μ, and surface μ. Conceptual views of these three types of muons are shown in Figure 2.4. A more detailed explanation is given later.

2.1.1 Continuous and pulsed muons

In the pulsed muon method, an instantaneously intense muon pulse, without identification of the arrival of each individual muon, is stopped inside the target material which is the object of research and the muon-related observables such as decay e^+/e^-, muonic X-ray, etc. are detected with reference to the arrival of the beam pulses by segmented detection methods. In the continuous muon method, after identifying the

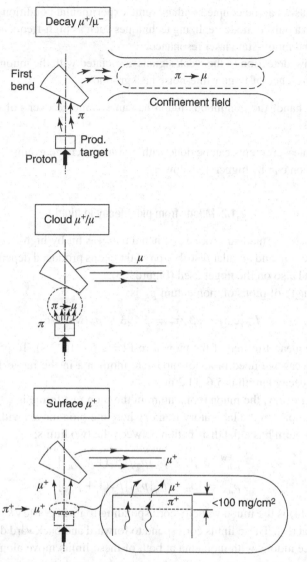

Figure 2.4 Three types of muon source (surface μ^+, cloud $\mu^\pm$, and decay $\mu^\pm$) produced by energetic protons from accelerators.

arrival of each muon, all the muon-associated events are recorded for each individual muon.

A number of advantages of the pulsed techniques over the continuous muon method have come to light. These can be summarized as follows:

1. Muon decay events, essential in both muon spin rotation/relaxation/resonance (μSR) and other types of experiment, can be measured out over a long time range and in a rate-unlimited manner.

2. The muon pulses can be coupled with extreme experimental conditions which can only be realized in a pulsed mode, realizing techniques such as high-frequency muon spin r.f. resonance, and muon-state laser resonance.
3. Phase-sensitive detection of the weak signals associated with the muon can be achieved even in the presence of large white-noise backgrounds.

On the other hand, the continuous muon has an advantage over pulsed muons in the following points:

1. Correlation measurements can be done with muon-associated events.
2. Time resolution can be higher, to below ns.

2.1.2 Muons from pion decay in flight

Pions are produced by nuclear processes when a target is hit by high-energy protons. The momentum spectrum and angular distribution of the pions produced depend on the primary beam energy and also on the target used (Figure 2.3).

The decay length of pions of momentum p_π is:

$$L_\pi(\text{cm}) = c\beta\gamma\tau_\pi = 5.593 \times p_\pi(\text{MeV/c})$$

where τ_π is the mean lifetime of the pion at rest ($= 2.6 \times 10^{-8}$ s). To produce muons by in-flight pion decay we need pions of moderate momenta in the range 100–200 MeV/c, where the pion decay length is 5.6–11.2 m.

In the $\pi \to \mu$ decay, the muon momentum in the pion rest frame is 29.8 MeV/c and its direction is isotropic. In the laboratory frame, where the pion moves with momentum p_π, the muon momentum has a flat distribution between the two limits:

$$p_\mu^{\text{Fw}} = (\beta_\pi + \beta_\mu^*)p_\pi/[\beta_\pi(1 + \beta_\mu^*)]$$

$$p_\mu^{\text{Bw}} = |\beta_\pi - \beta_\mu^*|p_\pi/[\beta_\pi(1 + \beta_\mu^*)]$$

where β_μ^* (0.2714) is the muon velocity corresponding to its 28.9 MeV/c momentum from the pion decay at rest. These limits correspond to forward and backward decays in the pion rest frame. Since muons with momenta at both of these limits move along the initial pion direction, they are easy to transport. Furthermore, they have definite polarizations of $+1$ and -1, respectively. Of particular interest to us are the backward muons, because their momentum (about half of p_π) is far from the initial beam momentum and so they can be cleanly separated from other particles such as π or e by bending magnets. In addition, muons of lower momentum have a higher stopping density. The muon momentum distribution for a given pion momentum and the corresponding decay angle in the laboratory frame are shown schematically in Figure 2.5.

To obtain high polarization we have to optimize several kinematic conditions for the pion decay in flight. This procedure involves: (1) selection of initial pion momentum p_π; (2) decay of pions into muons with minimal loss of π and μ; and (3) selection of suitable muon momenta. While selection of sharply defined momenta of p_π and p_μ is preferable

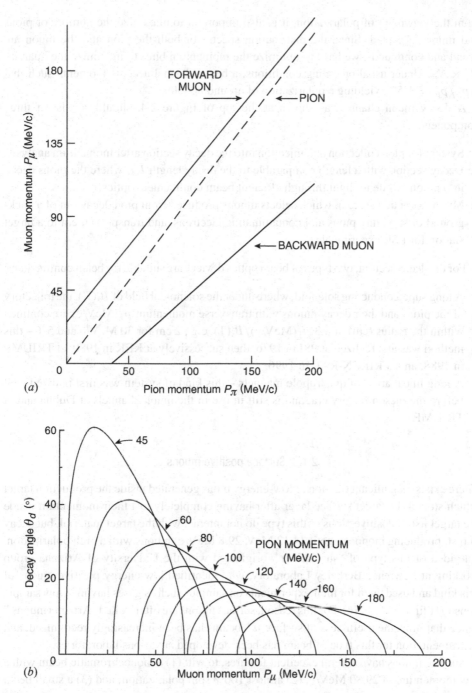

Figure 2.5 Summary of properties of decay muons; (a) momentum distribution and (b) decay-cone aperture angle for muons produced in pion decay in flight.

from the viewpoint of polarization, it is also important to maximize the number of pions and muons accepted. Since the distribution spectra of both the pion and the muon are broad and continuous, we have to optimize the momentum bites to maximize the quantity $P_\mu^2 \times N_\mu$. Under usual operating conditions, a beam channel accepts momenta such that $\Delta p_\mu / p_\mu = \pm 5\%$, yielding a polarization of around 80–90%.

A decay muon channel, as shown at the top of Figure 2.4, should consist of three components:

1. System for pion collection and injection into the decay section after momentum analysis.
2. Decay section with a length comparable to the decay length L_π, where the pions decay into muons while in flight through efficient beam confinement optics.
3. Muon extraction system, which selects muons produced from pion decay out of a background of surviving pions and contaminating electrons, and transports them to a target station for µSR experiments.

For the decay section, two types of beam optics devices are suitable for beam confinement:

1. A long superconducting solenoid, where under the solenoidal field of $B_s(T)$, the trajectory of the pions and their decay muons with transverse momentum p_T(MeV/c) are confined within the radius (cm) of 0.3 p_T(MeV/c)/B_s(T); e.g., 2 cm for 30 MeV/c and 5 T – this method was first realized at PSI in 1976, then successively at KEK in 1980, at TRIUMF in 1988, and at RIKEN-RAL in 1996.
2. A long linear array of quadrupole magnets – this kind of system was first realized well before the meson factory era, and is still in use in the muon channels at Dubna and at TRIUMF.

2.1.3 Surface positive muons

There exists a significant fraction of low-energy pions generated inside the production target which stop at the target surface layer after having completely lost their momentum inside the target itself. Positive pions of this type do not interact with the target material, but decay at rest, producing monoenergetic (4.12 MeV, 29.8 MeV/c) muons with a high polarization. The idea of this type of "surface" μ^+ originated with the University of Arizona group working at Lawrence Berkeley Laboratory, who identified low-energy positive muons of this kind and used them for their experiments on samples such as gases having a low sample density (Pfifer et al., 1976). For this reason, surface muons are often called "Arizona muons." Since that time, the usefulness of surface μ^+ beams has been increasingly recognized, and instrumentation for this type of beam has been developed at various laboratories.

Surface muons have several excellent features, to wit: (1) a monochromatic beam with a low momentum of 29.80 MeV/c; (2) almost 100% spin polarization; and (3) a small beam spot at the experimental target.

One problem of the surface μ^+ beam, however, is the existence of a large contamination of e$^+$, which have their origin mostly in $\pi^0 \rightarrow 2\gamma$ and $\pi^+ \rightarrow \mu^+ \rightarrow e^+$ decay inside the production target; the former is prompt and the latter is delayed with reference to the time of pion production. The e^+ contamination can be removed by installing a d.c. electrostatic separator at a suitable part of the beam channel. A d.c. separator with vertical electric field serves not only to remove contamination electrons but also, when combined with a crossed magnetic field, to rotate the spin polarization direction of the muon beam.

With a good target configuration one can obtain a μ^+ intensity as high as $10^5 \mu^+$ per 1 μA of protons. The surface μ^+ yield can be estimated based upon the positive-pion production cross-section (Tschalaer, 1978; Kettle, 1982). These muons have a range of about 160 mg/cm^2, and can be stopped within a layer of 20 mg/cm^2 thickness. This high stopping luminosity is a great advantage compared to the decay muon beam, and permits μ^+SR or muonium spin rotation/relaxation/resonance (MuSR) experiments on thin targets. When the beam channel is tuned to a momentum lower than 29.8 MeV/c, muons with even lower momentum can be obtained. Such "subsurface" muons are generated from pions decaying at some layer inside the target skin. The use of subsurface muons allows experiments with thinner than usual targets, as well as surface studies on conventional targets, but usually the intensity is lower and the polarization somewhat reduced compared to surface μ^+.

There is no possibility of a surface μ^- beam because prompt nuclear capture of the π^- takes place inside the target before π^- decay has time to occur, except for a small probability in the case of liquid H_2 or He (Bowen, 1985).

2.1.4 Cloud muons

There is also some fraction of pions which decay in flight in the free space close to the production target. Suppose we have a pion beam channel with a bending magnet at the forward end to act as a momentum analyzer. The muons produced between the production target and the bending magnet take the form of a cloud around the target and can be transported together with the pions, as seen in Figure 2.4 (Tanabe, 1971). Since the momenta of the parent pions are unknown, these muons do not satisfy any kinematic conditions to assure their polarization; they are a mixture of forward- and backward-decay muons. Therefore, we expect their net polarization to be small. However, since the pion number spectrum at the production stage has a steep increase with p_π, cloud muons having momenta in the distribution's tail region include more forward-decay muons than backward-decay muons. Thus, a significant polarization for low-momentum cloud muons is assured.

2.1.5 Beam optics components for MeV muons

The beam optics, otherwise known as the muon channel, for an accelerator producing MeV muons should comprise the following components: a pion production target, an initial momentum-selecting magnet, a decay section for the decay μ^- and/or electrostatic optics for background elimination, and a second momentum-selecting magnet followed by a focusing magnet to direct the muons produced on to a target sample.

Since primary protons from the accelerator can generate secondary particles such as neutrons and radioactive nuclei not only at the pion production target but also at the surrounding beam line components, including the beam dump, radiation safety has to be thoroughly considered, and all the elements should be installed inside compact radiation shielding of sufficient thickness and density. This precaution is particularly important when the muon beam line and associated experimental facility are installed as a tandem style in front of a spallation neutron source and facility.

Advanced generation of decay muons became possible after realization of a large-scale superconducting solenoid with parameters such as a 50 kG field, 6–8 m long and a 10–15 cm inner bore. Details of the solenoid design as well as the refrigeration system are described in the references (Vecsey, 1975; Nagamine, 1981).

Except for backward muons generated from pion decay in flight, the separation of muons from contaminating particles such as e^- or e^+ is necessary to obtain a muon beam of high purity. After selecting the beam momentum p_μ using magnetic field optics, the application of an electric field E_0 perpendicular to the beam direction with a fixed field length L causes a deflection which depends on the particle mass; particles having different velocities experience the action of the field for different lengths of time. When a magnetic field B_0 is applied perpendicular to E_0 (crossed fields), the following condition is necessary for a particle with velocity β_0 ($= v/c$) to traverse the field region without deflection: $\beta_0 = AE_0/B_0$, with $A = 3.33$ G (kV/cm). Thus, by adjusting E_0 and B_0, particles with velocity of the corresponding value of β_0 are selected and transmitted under an undeflected manner.

The spin of a polarized muon, during nondeflective passage through crossed fields, undergoes precession through an angle ϕ; $\phi = eB_0L/P_\mu\gamma_0$. Thus, a muon which is originally longitudinally polarized can become transversely polarized by selecting L and B_0 appropriate to p_μ. For typical surface μ^+ with $p_\mu = 27$ MeV/c ($\beta_0 = 0.247$ and $\gamma_0 = 1.032$), we need (for example) B_0 of 496 G with $L = 2.94$ m and E_0 of 36.8 kV/cm (368 kV for 10 cm gap of the separator) to yield 90° spin rotation (Beveridge et al., 1985).

2.2 eV–keV slow muons

As described above, muons are usually obtained either from pion decay in flight (decay μ^+, μ^-) or from π^+ decay at the surface skin of a pion production target (surface μ^+). In these cases, even with a carefully designed beam channel, the width range of the stopping muons is larger than 10 mg/cm^2. A significant intensity decrease is thus inevitable when we wish to study samples with thinner than this range. If we could establish a method to increase the slow μ^+ intensity, many new types of muon science experiments would become possible.

To date, there have been several proposals concerning how to achieve such slow muon beam production. Representative ideas can be summarized as follows in order of the lowest energy attainable. As summarized in Table 2.1, the proposed methods are significantly different depending upon slow μ^+ or slow μ^- and on how slow it is:

1. 0.2 eV ∼, μ^+: the thermal muonium ionization method adopts the phenomenon of thermal Mu emission which occurs when conventional (MeV) muons are stopped inside specially selected materials, followed by ionization of the Mu (Nagamine and Mills, 1986).

Table 2.1 Proposed and realized[a] muon cooling method

Energy	Method	Type of Muon
1000 MeV	Ionization cooling	μ^+, μ^-
10 MeV	Prism	μ^+, μ^-
1 MeV	Phase space compression	μ^+, μ^-
	μ^- Reemission from μCF	μ^-
1 keV	Frictional cooling (PSI)[a]	μ^+, μ^-
	Cold moderator (TRIUMF/PSI)[a]	μ^+
1 eV	Thermal muonium ionization (KEK)[a]	μ^+

μCF, muon catalyzed fusion; PSI, Paul Scherrer Institute; TRIUMF, Tri-University Meson Facility; KEK, High Energy Accelerator Research Organization.

2. 2 eV $\sim$, μ^+: the cold moderator method adopts a stopping material with a large energy gap, such as a solid layer of a rare-gas element (Ne, Ar, etc.) to slow conventional muons without Mu formation (Harshman et al., 1986).

3. A few keV, μ^+/μ^-: the frictional cooling method uses a positively changing character of the energy loss (dE/dx) of the muon against energy (E).

4. A few keV, μ^+/μ^-: the beam-cooling method uses electromagnetic confinement techniques to narrow the phase space of each injected conventional muon by acceleration or deceleration by applied electric fields (Taqqu, 1986).

5. $\sim$ 10 keV, μ^-: the muon catalyzed fusion (μCF) leads to a generation of slow μ^- which uses the process of μCF (Nagamine, 1989).

6. 100 keV $\sim$ MeV, μ^+/μ^-: the inverse cyclotron method involves the slowing-down of a high-energy (> MeV) muon beam under an applied transverse magnetic field in the presence of a gas introduced for energy loss (Simmons, 1990).

7. MeV $\sim$ 100 MeV, μ^+/μ^-: in the Phase Rotated Intense Slow Muon Source (PRISM) method (Kuno, 2000), using phase-space ($\Delta E - \Delta t$) rotation due to an applied radiofrequency wave, the cooling in ΔE is realized by expanding Δt.

8. $\geq$200 MeV, μ^+/μ^-: ionization cooling again uses the positively changing character of the energy loss of the muon against energy.

Details of the first five methods listed above are given in the remainder of this chapter.

2.2.1 Thermal Mu and the laser resonant ionization method for slow μ^+ generation

The method of laser resonant ionization of thermal Mu emitted from a hot noble metal target such as tungsten was motivated by the discovery of thermal Mu emission into vacuum from the surface of hot tungsten (W) in experiments conducted at the High Energy Accelerator Research Organization–Meson Science Laboratory (KEK-MSL) (Mills et al., 1986).

Systematic studies have been carried out for most of the noble metals (Matsushita *et al.*, 1998). The overall result is consistent with the old measurements of solution enthalpy of hydrogen adsorption into the metals (Frauenfelder, 1969); the more positive the solution enthalpy (the more unstable hydrogen inside the material) gives us the larger thermal Mu yield. The idea was encouraged by the laser-resonant excitation experiments, also conducted at KEK-MSL (Chu *et al.*, 1988), as described in section 1.3.

Thermal muonium can also be produced by stopping energetic μ^+ in a powdered form of a material such as SiO_2, where production of muonium can be expected inside its bulk form. There, the following picture can be expected for the production of thermal Mu in a vacuum. The μ^+, after slowing down during a passage through the powder ensemble, stops inside a powder particle which has a diameter of 10 nm (100 Å). There, a large fraction of μ^+ changes to Mu, as expected from the process which will be described in Chapter 3. Then, Mu produced at thermal energy can take a thermal diffusion towards the boundary of the powder particle. There, a competing process exists at the boundary; namely, a reflection back to the powder interior or a penetration towards the vacuum space in between the powder particles. The penetrated Mu may have a successive collision process with powder particles, where another competing process exists, namely, reabsorption into the powder or scattering from the powder surface to the vacuum. Where the situation in these two competing processes is favorable, as is known to be the case for some specific materials, a substantial fraction of thermal Mu is produced in vacuum. The SiO_2 powder is known to be one of the materials which provides these favorable situations (Beer *et al.*, 1985).

The original experiment of slow μ^+ generation by the thermal Mu laser ionization was conducted at KEK-MSL (Nagamine *et al.*, 1995) by placing a thermal Mu-producing material (50 μm thick W at 2300 K) at a 500 MeV proton beam line right behind π^+-producing material (2 mm thick boron nitride), because of the intensity required. All the set-up was placed at a pressure below 10^{-8} mm Hg (10^{-8} Torr). Thus, as can be seen in the inset of Figure 2.6, the intended scenario is the following: low-energy π^+ produced in the boron nitride stops in the hot W where, after conversion of π^+ to μ^+ and deceleration and diffusion of the μ^+ inside the W foil, the thermal Mu is emitted from the W surface.

For the ionization process, a vacuum ultraviolet (VUV) (122.09 nm) light pulse of a few μJ/(5 ns) intensity yielding the initial 1s–2p transition is generated synchronously with the proton beam pulse (using a signal from the proton accelerator), and goes directly into the target region. The second laser of 355 nm, with an intensity of 30 mJ/(5 ns), is introduced into the target region to ionize the 2p state of the Mu.

During the course of the development of the ultraslow μ^+ production method, the following experimental findings were obtained (Miyake *et al.*, 1997):

1. The activation energy of the production process was determined by measuring the yield of ultraslow μ^+ as a function of temperature; resulting activation energies proved to be mass-independent among μ^+, H, D, and T, suggesting a thermionic emission mechanism of the hydrogen-like neutral particles from the hot W surface.
2. The ultraslow μ^+ yield, at this stage of the project development, is 10 μ^+ per second per 1 μA of 500 MeV protons and 10 μJ/pulse of 112 nm laser power.

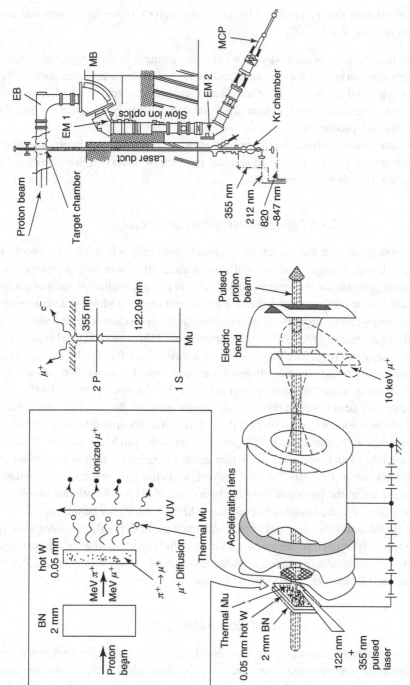

Figure 2.6 Concept of ultraslow positive muon production by resonant laser ionization of thermal muonium from hot tungsten (W) placed in the primary proton beam line. BN, boron nitrate; VUV, vacuum ultraviolet; MCP, multichannel plate.

3. The polarization of the μ^+ produced by a spin-unresolved $1s \rightarrow 2p \rightarrow$ unbound laser ionization was found to be 50%.

Assuming that the present method is free from any primary beam-associated radiation effects at higher proton intensities, we can expect a straightforward increase in the ultraslow μ^+ intensity (up to $10^6/s$) as a result of installing the present system at a higher-intensity pulsed proton source. This scaling feature is a significant advantage compared to the other methods of slow μ^+ production.

An alternative idea to increase the intensity of slow μ^+ is to employ a large-acceptance, very-high-intensity μ^+ generator and to stop the production of very-high-intensity μ^+ in the hot tungsten. Details will be given later.

2.2.2 Cold moderator method for slow μ^+

During its passage through the matter, the originally energetic muon loses its energy by transferring its kinetic energy to the material via electric excitation or ionization processes. When an energy gap exists in the material for any kind of electric excitation, such an energy-loss process terminates at the energy below the gap so that the μ^+ with lower kinetic energy than the gap energy can go through the material and may have a chance to be emitted from the material surface as an epithermal μ^+ (Harshman et al., 1986). Since several cryocrystals have a finite energy gap (e.g., 22 eV for Ne, 14 eV for Ar, 15 eV for N_2), by placing a cold solidified gas target as the stopping material, one can expect a generation of slow (eV $\sim$ 10eV)μ^+ beam by using the set-up shown in Figure 2.7 (Morenzoni et al., 1995).

There are several advantages in this cold moderator method, in particular, compared to the thermal Mu laser ionization method: (1) the experimental set-up is relatively simple and produce more than 100 slow μ^+/s; and (2) polarization of the produced slow μ^+ is almost 100%. On the other hand, some disadvantages exist: (1) conversion efficiency from one MeV μ^+ to slow μ^+ is 10^{-5}–10^{-6} so that very-high-intensity μ^+ injection is inevitable; (2) the phase space of the produced slow μ^+ is not small; (3) for the production of very-high-intensity slow μ^+, the heating effect on the cold moderator may be inevitable.

Utilizing the advantageous features of the moderator method, highly polarized slow μ^+ has been extensively applied to the muon spin rotation/relaxation studies of thin materials such as high-T_c superconductor surfaces (Niedermayer et al., 1999; Jackson et al., 2000a), iron nanoclusters (Jackson et al., 2000b), and so on.

2.2.3 Frictional cooling

As described in Chapter 3, the energy-loss process of energetic muons takes a systematic way depending upon the energy range. There are energy regions which show a positive energy derivative of stopping power with respect to energy ($dS/dE > 0$), both below 100 keV and above 200 MeV. There, one can expect an increase of stopping power by increasing the energy, thus producing a mechanism of cooling – energy loss during passage through material followed by longitudinal energy restoration by acceleration with applied electric

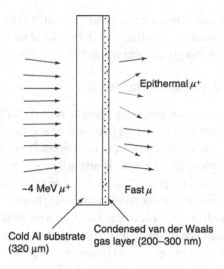

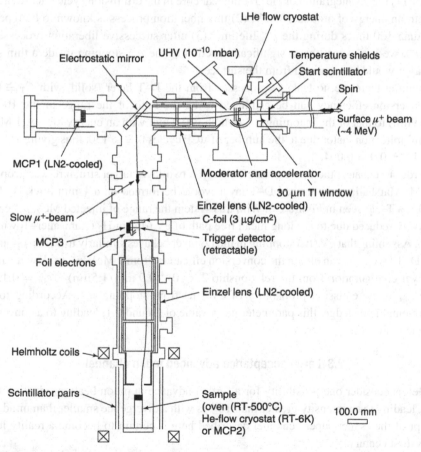

Figure 2.7 Slow μ^+ generation by the cold moderator method; (a) concept and (b) realized set-up at Paul Scherrer Institute. LN_2, liquid nitrogen; MCP3, multichannel plane no. 3, LHe, liquid helium; UHV, ultrahigh vacuum. Reproduced from Morenzoni (1999).

field. The lower-energy region is called frictional cooling (Daniel, 1989) and the higher-energy region is called ionization cooling (Neuffer, 1994). Frictional cooling was applied to energy loss measurements of low-energy μ^+ and μ^- at PSI.

2.2.4 μCF method for slow μ^-

In the case of negative muons (μ^-), it has long been thought very difficult to produce an intense slow μ^- beam for the following reasons: (1) because of the strong absorption of stopped π^- inside matter, the π^- to μ^- decay process is completely suppressed inside a finite bulk of target material, and consequently there is no surface μ^- production; (2) because of muonic atom formation, the stopped μ^- cannot be liberated from a condensed matter target after thermalization, and thus no reemission can be expected in the case of μ^-.

In order to overcome the second difficulty, a new idea exists for a potential source of slow μ^-, which avails itself of the assistance of the muon catalyzed fusion (μCF) phenomena which is described in Chapter 5, as seen in Figure 2.8 (Nagamine, 1989). The principle is as follows: (1) the disintegration of the ^{5}He nuclear core in the dtμ fusion cycle releases a slow μ^- with an energy of around 10 keV; (2) this liberation process is known to be repeated more than 100 times during the μ^- lifetime; (3) after successive liberation processes of slow μ^-, we can expect that a significant fraction of the μ^- stopping inside a thin solid D–T layer would be reemitted from the surface.

Assuming there are no leakage processes from the D–T layer (solid, with $C_T \cong 0.3$), the conversion efficiency can be estimated to be the ratio of the range of the 10 keV μ^- (0.3 μm) to that of the incoming μ^- (which is 0.9 mm with an energy of, say, 1 MeV). The multiplication factor due to the number of μCF cycles is $\cong \sqrt{150}$, thus giving $\sqrt{150} \times 0.3 \times 10^{-3}/0.9 = 0.004$.

In order to enhance the conversion efficiency, a two-layer target structure was proposed by G.M. Marshall; an optimized D–T layer would be formed on a 1-mm-thick H$_2$ layer with 0.1% T$_2$, as seen in Figure 2.8. With this system the range of injected MeV μ^- can be effectively reduced due to the long mean free path of eV (dμ) in H$_2$ (Ramsauer–Townsend effect). Assuming that eV (tμ) stopping in the D$_2$ layer behaves similarly to eV (tμ) stopping in the D–T layer, one can obtain the conversion efficiency from 1 MeV μ^- to slow μ^- in the two-layer configuration from the relationship $2 \times 10^{-4}(0.9 \text{ mm}/1.5 \mu\text{m}) \times \varepsilon_{t\mu} = 0.12\varepsilon_{t\mu}$, where $\varepsilon_{t\mu}$ is the emission probability of eV (tμ) from stopping μ^-. According to our experimental knowledge, this parameter has a value of around 0.1, leading to a conversion efficiency of 0.012.

2.3 Large acceptance advanced muon channel

Now let us consider one possibility for a further advance in muon beam production technique, leading to an intensity greater than 10^{10}/s with a divergence smaller than mrad. The concept of the "super-super" channel described here is certain to become a reality in the twenty-first century.

In all the existing muon channels at major accelerator facilities, the front-end collection optics for either pions or surface μ^+ is placed directly adjacent to the production target in

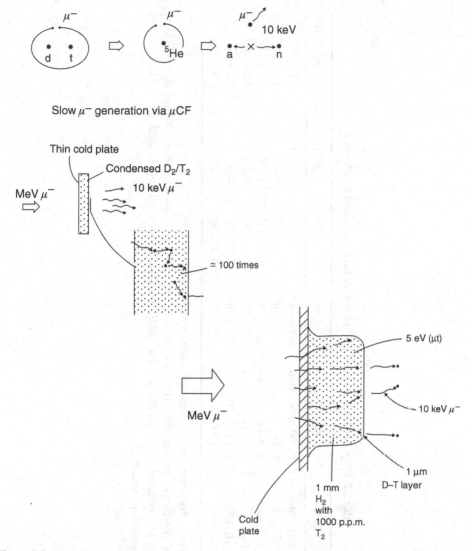

Figure 2.8 Concept of slow (keV-region) negative muon production using the muon catalyzed fusion (μCF) process in a thin D-T layer.

the primary proton beam, with the acceptance dependent solely upon the geometrical solid angle determined by the production target location and the aperture of the front-end optics element. Consequently, the realistic upper limit for the acceptance is 100 mrad at most.

To gain a drastic improvement over the present situation, several ideas have been proposed, as summarized in Table 2.2. The common feature shared by these proposals is the placement of a large superconducting magnetic field right at the production target in order to collect the largest possible fraction of the pions produced.

Recently, there has been considerable progress in design work related to the front-end optics for neutrino-factory and $\mu^+\mu^-$ colliders (Raja *et al.*, 2001), and also in work related to

Table 2.2 Proposed high-intensity μ^- source at various hadron accelerators

Project	Institute	Accelerator	p, d/s	μ^-/s	μ^-/p,d	μ^-/power(p,d)(MW)
LFNC[a]	INR – Moscow	p, 500 MeV, 100 μA	6.2×10^{14}	10^{11}	1.6×10^{-14}	2×10^{12}
LFNC[b]	AGS – BNL	p, 24 GeV, 3 μA	1.9×10^{13}	4×10^{11}	0.021	5.6×10^{14}
$\mu^+\mu^-$ collider[c]	BNL etc.	p, 30 GeV, 0.25 μA	$2.3 \times 10^{13}/15$	$4 \times 10^{12}/15$	0.17	3.6×10^{13}
μCF n-source[d]	PSI	d, 1.5 GeV, 12 mA	7.5×10^{16}	10^{15}	0.013	5.6×10^{13}
General purpose[e]	JHF-KEK	p, 3 GeV, 200 μA	1.3×10^{15}	1.3×10^{13}	0.01	2.2×10^{13}
μCF reactor[f]	Gatchina	d, 1.5 GeV, 12 mA	7.5×10^{16}	1.5×10^{16}	0.20	8.0×10^{14}

LFNC, Search for Lepton Flavor Non-conservation; μCF, muon catalyzed fusion.

[a] Abadjev, V.S. et al. (1992). INR preprint 786; [b] Molzon W. et al. (1996). *UCI Phys. Tech. Rep.* 96-30; [c] Palmer, R. et al. (1996). *Nucl. Phys.*, **B**, 51A 61; [d] Petitjean, C. (1993). *PSI Rep.* PR-93-09; [e] Nagamine, K. et al. (1996). *UT-MSL Internal Rep.*; [f] Petrov, Yu. V. (1982) *Atomkerm-Kerntech.*, 46 25.

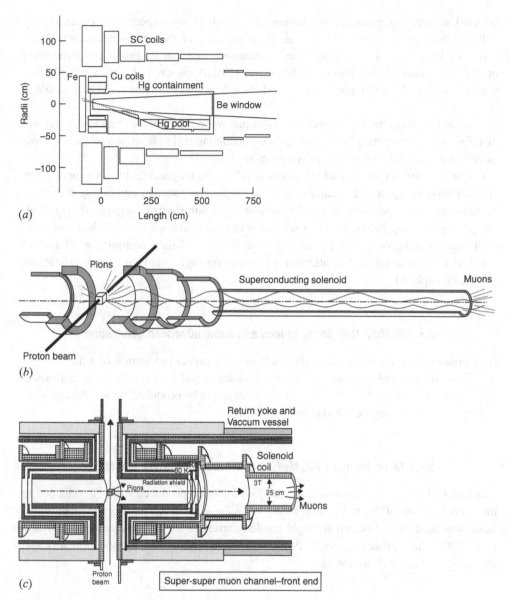

Figure 2.9 (a) Target and solenoid configuration for the pion capture section of the proposed neutrino factory (Raja *et al.*, 2001). (b) Conceptual view and (c) details of the target area of the proposed super-super muon channel (Ishida and Nagamine, 1998).

pion collection for muon rare decay experiments. With a view towards future projects at next-generation muon-science experimental facilities, a practical design for such a "super-super" muon channel has been proposed (Ishida and Nagamine, 1998). Some features of the proposed design can be seen in Figure 2.9. In particular, the 90° design, with the pion extraction axis perpendicular to the proton beam axis, was selected in order to minimize irradiation effects which are expected to occur mostly collinear with the proton beam. It was the conclusion of the optics design work that a μ^+/μ^- beam with a reasonably small momentum bite can be produced at the level of $10^{11}\mu$/s either at the upgraded ISIS–Rutherford Appleton Laboratory (ISIS-RAL: 0.8 GeV $\times$ 300 μA) or at Japan Proton Accelerator Research Complex (J-PARC: 3 GeV $\times$ 200 μA).

The pilot model of the super-super muon channel was realized at the Dai-Omega channel of KEK, where, using an axial-focusing superconducting coil system, the surface μ^+ was extracted in the solid angle of 1 str (Miyadera *et al.*, 2002).

Given the existence of such a high-intensity μ^+/μ^- beam generated from a super-super channel, one can expect high-quality slow μ^+/μ^- beam generation by combining the technical features of the super-super channel with those of the ultraslow muon generation method. For μ^+, by stopping intense incident μ^+ in hot tungsten sheets, a slow μ^+ beam of quite high intensity and good quality can be generated. However, there is no similar good method for slow μ^- generation, and so alternative methods for high-quality μ^- beam production need to be explored.

2.4 100 MeV–GeV decay muons and some advanced generation

By increasing pion energy beyond 100 MeV one can expect production of a high-energy μ^+ beam. By extrapolating the formula for forward μ and backward μ, one can expect $p_\mu^{\mathrm{FWD}} \cong 1.1 p_\pi$ and $p_\mu^{\mathrm{BWD}} \cong 0.57 p_\pi$. Also, decay μ can be produced by $K^\pm$ decay: some distinguished examples are listed below.

2.4.1 Monochromatic 230 MeVμ^+ by K^+ decay at production target

Just like surface μ^+, a monochromatic μ^+ beam can be produced from K^+ stopped at the surface region of the production target via $K^+ \rightarrow \mu^+ + \nu_\mu$. Such a monochromatic beam was used for the search for right-handed current in the weak interaction (Hayano *et al.*, 1984; Imazato *et al.*, 1992). Possible advanced muon beam production was also emphasized (Tanaka *et al.*, 1992).

2.4.2 100 GeV muon beam for the EMC experiment

By using the 450 GeV Super Proton Synchrotron at the European Organization for Nuclear Research (CERN SPS), high-energy pions are produced to provide high-energy muons in the energy range between 100 and 280 GeV. These high-energy muons in the European Muon Collaboration (EMC) experiment have been used for deep inelastic scattering from nuclei to probe charge and spin structure at very high momentum transfer (Alkofer *et al.*, 1985; Voss, 1992).

2.5 GeV–TeV cosmic-ray muons

Primary cosmic rays interact with nuclei in the atmospheric air and produce secondary cosmic rays through high-energy reactions (Adair and Kasha, 1976). The primary cosmic rays consist largely of proton (95%) and He nuclei (5%), with a small contribution from heavier nuclei up to Fe. The energy spectrum of primary cosmic rays follows approximately the relationship $N(E) \propto 1.1 \times [E(\text{GeV})]^{-1.75}$ at high energies, with a leveling-off below a few GeV.

In these interactions of primary cosmic-ray protons with nuclei like N, O, and others in the air, a large quantity of pions and kaons are produced. Charged pions and kaons decay into muons, while neutral pions decay into 2γ. Therefore, most of the secondary cosmic rays observed at the earth's surface are muons (70%) and electrons (30%).

Here we would like to summarize the properties of the cosmic-ray muons produced by the decay of pions and kaons generated by nuclear reactions between primary cosmic rays and atmospheric air. The energy spectrum of the cosmic-ray muon was measured experimentally as a function of the zenith angle θ_z, and the results are summarized in Figure 2.10. As can be seen from this figure, the intensity is more pronounced at $\theta_z = 0$ below 100 GeV, while it is larger at $\theta_z = 90°$ above 200 GeV.

This tendency can be understood with the aid of the following argument. According to a compilation of the existing data (Adair and Kasha, 1976), the intensity increase $I_0(\theta_z, E)$ as a function of energy for high-energy muons arriving with zenith angle θ_z can be written

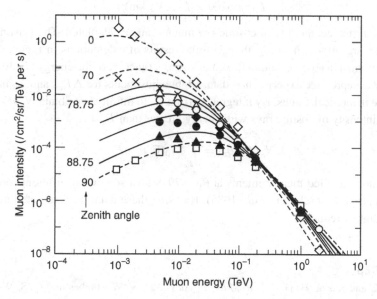

Figure 2.10 Energy spectrum of cosmic-ray muons arriving at the earth's surface with various zenith angles. Experimental data are taken from those cited in Adair and Kasha (1976). The curves are the results of the model calculation described in the text.

according to the following formula:

$$I_0(E, \theta_z) = \mathrm{d}N/\mathrm{d}E$$
$$= 1.2 \times 10^{-6} E(\mathrm{TeV})^{-2.7}$$
$$\times [0.9/(1 + E\cos\theta_z/E_\pi) + 0.1/(1 + E\cos\theta_z/E_k)]$$
$$(\mathrm{s}^{-1}/\mathrm{cm}^2/\mathrm{sr}/\mathrm{TeV})$$

where $E_\pi = 0.092$ TeV and $E_k = 0.54$ TeV.

Since a layer of air of thickness $L_0 = 1.013$ kg/cm^2 has a corresponding energy loss $\Delta E_{\mathrm{air}} = 0.0026$ TeV, a cosmic-ray muon passing through the air with a zenith angle θ_z has an energy loss of $\Delta E_{\mathrm{air}}/\cos\theta_z$. Thus, the intensity of cosmic-ray muons with final energy E is determined by the larger energy quantity E' which is the sum of E and $\Delta E_{\mathrm{air}}/\cos\theta_z$. Thus, the energy spectrum on the surface, $I(E, \theta_z)$, is determined by $I_0(E', \theta_z)$ with the required correction.

$$I(E, \theta_z) = \mathrm{d}N(E', \theta_z)/\mathrm{d}E', \text{ with } E' = E + \Delta E_{\mathrm{air}}/\cos\theta_z.$$

For θ_z near $90°$, a correction for the spherical nature of the earth is needed so that:

$$L/L_0 = [(a^2\cos^2\theta_z + 2ab + b^2)^{1/2} - a\cos\theta_z]/b$$

where a is the earth's radius and b is the air thickness around the earth. In addition, the muon intensity loss due to in-flight muon decay should be taken into account, where the decay length (L_d) of a muon with energy E is given as:

$$L_d = 6200 \times E(\mathrm{TeV})(\mathrm{km})$$

Using these formulae, the flux of cosmic-ray muons can be calculated. The existing experimental data are summarized together with the results of calculations in Figure 2.10; the data are presented in terms of muon flux intensity as a function of the energy (E) for various θ_z. In order to reproduce experimental data, slight adjustments for ΔE_{air} and L_d have been made to the theoretical values. By integrating from E_c to infinity, we obtain $N_\mu(E_c, \theta_z)$, the integrated intensity of cosmic rays with energy greater than E_c, i.e.:

$$N_\mu(E_c, \theta_z) = \int_{E_c}^{\infty} I_0(E, \theta_z)\mathrm{d}E$$

By using more detailed measurements at $\theta_z = 79$–$90°$, a somewhat different formula for $I(E, \theta_z)$ is proposed (Alkofer et al., 1985). By using these data, the integrated muon flux becomes slightly reduced.

REFERENCES

Adair, R. K. and Kasha, H. (1976). In *Muon Physics I*, ed. V. W. Hughes and C. S. Wu, p. 323. Academic Press.

Alkofer, O.C. et al. (1985). *Nucl. Phys.,* **B259**, 1.

Beer, G. et al. (1995). *Phys. Rev. Lett.,* **57**, 671.

Beveridge, J. L. et al. (1985). *Nucl. Instr.,* **A240**, 316.

Bowen, T. (1985). *Physics Today*, **38**, 22.

Chu, S. *et al.* (1988). *Phys. Rev. Lett.*, **60**, 101.

Daniel, H. (1989). *Muon Catal. Fusion*, **4**, 425.

Eaton, G. H. *et al.* (1988). *Nucl. Instr.*, **A269**, 483.

Frauenfelder, R. (1969). *J. Vac. Sci. Tech.*, **6**, 388.

Harshman, D. R. *et al.* (1986). *Phys. Rev. Lett.*, **56**, 2850.

Hayano, R. S. *et al.* (1984). *Phys. Rev. Lett.*, **52**, 329.

Imazato, J. *et al.* (1992). *Phys. Rev. Lett.*, **69**, 877.

Ishida, K. and Nagamine, K. (1998). *KEK Proc.*, **98-5 II**, 12.

Jackson, T. J. *et al.* (2000a). *Phys. Rev. Lett.*, **84**, 4598.

Jackson, T. J. *et al.* (2000b). *J. Phys. Condens. Matter*, **12**, 1399.

Kettle, P.-R. (1982). *SIN Rep.*, **TM-31-20**.

Kuno, Y. (2000). *Nucl. Instrum. Meth.*, **A451**, 233.

Lindenbaum, S. J. and Sternheimer, R. M. (1957). *Phys. Rev.*, **105**, 1874.

Matsushita, A. *et al.* (1998). *Phsy. Lett.*, **A244**, 174.

Matsuzaki, T. *et al.* (2001). *Nucl. Instrum.*, **A465**, 365.

Mills, A. P. Jr *et al.* (1986). *Phys. Rev. Lett.*, **56**, 1463.

Miyadera, H. *et al.* (2001). *Hyperfine Interact.*, **138**, 505.

Miyake, Y. *et al.* (1997). *Hyperfine Interact.*, **106**, 237.

Morenzoni, E. (1999). In *Muon Science*, ed. S.L. Lee, S.H. Kilcoyne and R. Cywinski, p. 343. Edinburgh: SUSSP/Bristol: Institute of Physics Publishing.

Morenzoni, E. *et al.* (1995). *Phys. Rev. Lett.*, **72**, 2793.

Nagamine, K. (1981). *Hyperfine Interact.*, **8**, 787.

Nagamine, K. (1989). *Proc. Jpn Acad.*, **65B**, 225.

Nagamine, K. (1998). *AIP Conf. Procs.*, **441**, 295.

Nagamine, K. and Mills, A. P. Jr (1986). *Los Alamos Rep. N,* **LA- 10714C**, 216.

Nagamine, K. *et al.* (1981). *IEEE Transact. Magnetics,* **MAG 17**, 1882.

Nagamine, K. *et al.* (1995). *Phys. Rev. Lett.*, **74**, 4811.

Nagamine, K. *et al.* (1996). *Hyperfine Interact.*, **101/102**, 521.

Neuffer, D.V. (1994). *Nucl. Instr.*, **A350**, 27.

Niedermayer, Ch. *et al.* (1999). *Phys. Rev. Lett.*, **83**, 3932.

Pifer, A. E. *et al.* (1976). *Nucl. Instrum.*, **135**, 39.

Raja, R. *et al.* (2001). (ed.) *Fermilab Conf.*, *01-226-E*.

Simmons, F. (1990). *Exotic Atoms in Condensed Matter, Proceedings in Physics*, vol. 59, p. 33. Berlin: Springer.

Tanabe, K. (1971). *Particle Accelerators*, **2**, 211.

Tanaka, K. H. *et al.* (1992). *Nucl. Instr.*, **A316**, 134.

Taqqu, D. (1986). *Nucl. Instr.*, **A247**, 288.

Tschalaer, C. (1978). *LAMPF Rep.*, **LA-7222-MA**.

Vecsey, G. (1975). *SIN Rep.*, **PR-75**, 002.

Voss, R. (1992). *CERN-PPE* **9**, **2**, 45.

3

Muons inside condensed matter

Various types of muons, with energies ranging from eV to TeV (10^{12} eV), can be introduced into condensed matter where, after energy loss, they eventually come to a stop. Apart from the application of cosmic rays to the study of the internal structure of geophysical-scale objects (to be described in Chapter 9), most muon science studies deal with muons which have been stopped inside matter in this way. Therefore, it is important to consider the history of these muons from their introduction into matter at high energies (usually a few MeV) until the point where they reach thermal energies and stop.

3.1 Stopping muons in matter and polarization change

Let us consider either μ^+ or μ^-, which, after production by some conventional intermediate-energy accelerator, is introduced into condensed matter with an energy in the range from several MeV to several tens of MeV. We assume that the beam is fully polarized.

The time-dependent changes corresponding to the various energy loss processes for the μ^+ and the μ^- are summarized in Figures 3.1 and 3.2, where the changes of polarization in each case are also shown. One can see that below a few keV there is a significant difference between the μ^+ and the μ^-.

At energies higher than a few keV of velocity v, where there is not much difference between μ^+ and μ^-, the main mechanism of energy loss is ionization and the process is described by the Bethe formula excluding the density-effect correction:

$$-\frac{dE}{dx} = 4\pi N_a r_e^2 m_e c^2 \rho \frac{Z}{A} \frac{1}{\beta^2} \left[\ell n \left(\frac{2m_e \gamma^2 v^2}{I} \right) - \beta^2 \right]$$

where $r_e = 2.817 \times 10^{-13}$ m; m_e = electron mass;
N_a = Avogadro number; I = mean ionization potential;
$Z/A/\rho$ = charge/atomic weight/density of the absorbing material;
$\beta = v/c$; and $\gamma = 1/\sqrt{1-\beta^2}$.

The use of this formula enables us to obtain quantitative information regarding the range of the muon in condensed matter R_0 (length required for the complete loss of the muon's initial energy) versus initial energy/momentum. In the momentum (p) range from 20 to 80 MeV/c,

Positive muon

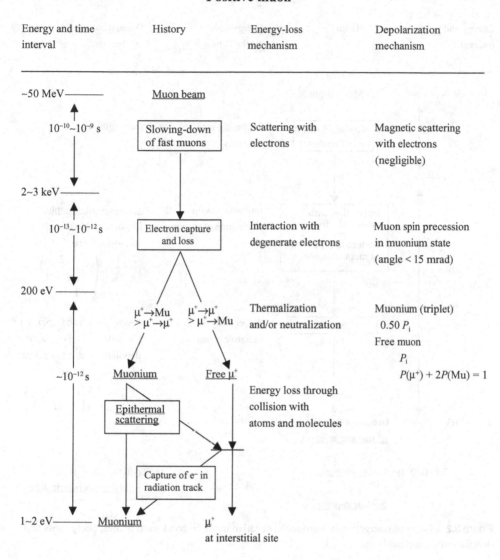

| Energy and time interval | History | Energy-loss mechanism | Depolarization mechanism |

~50 MeV ——— Muon beam

10^{-10}~10^{-9} s — Slowing-down of fast muons — Scattering with electrons — Magnetic scattering with electrons (negligible)

2~3 keV ———

10^{-13}~10^{-12} s — Electron capture and loss — Interaction with degenerate electrons — Muon spin precession in muonium state (angle < 15 mrad)

200 eV ———

$\mu^+ \to Mu > \mu^+ \to \mu^+$ $\mu^+ \to \mu^+ > \mu^+ \to Mu$ — Thermalization and/or neutralization — Muonium (triplet) 0.50 P_i
Free muon P_i

~10^{-12} s — Muonium   Free μ^+ — $P(\mu^+) + 2P(Mu) = 1$

Epithermal scattering — Energy loss through collision with atoms and molecules

Capture of e$^-$ in radiation track

1~2 eV ——— Muonium μ^+
at interstitial site

2.2 µs   Decay of muon

Figure 3.1 History of energetically introduced positive muon in condensed matter; energy loss and depolarization mechanisms.

R_0 is known to be proportional $p^{3.5}$. More detailed theoretical/experimental data are summarized in Figure 3.3 at higher energy and in Figure 3.4 below 100 keV with experimental data (Mühlbauer et al., 1999). The dE/dx decreases with increasing velocity in the ionization process and, due to the increase in radiative loss effects at higher energies, minimum

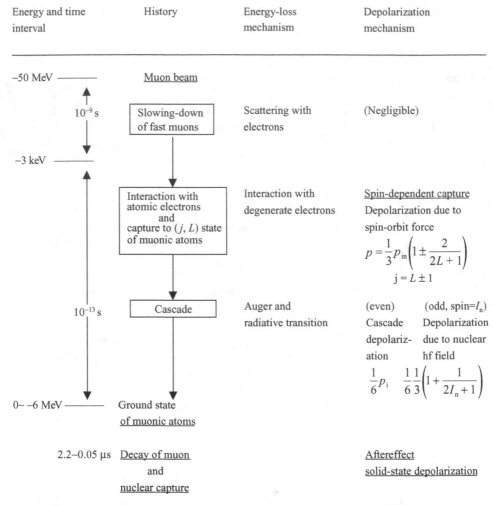

Figure 3.2 History of energetically introduced negative muon in condensed matter; energy loss and depolarization mechanisms.

ionization exists at around 200 MeV. On the other hand, below the peak at around 10 keV, muons become too slow to ionize the atoms. The importance of the sign of dE/dx has already been explained in relation to beam cooling (see section 2.2). The dE/dx data give a useful practical guide to the spatial distribution of muon stopping positions inside samples of condensed matter to be studied in muon science experiments.

During the slowing-down process via ionization, as a result of the statistical nature of the multiple collision processes, a sharply collimated initial muon beam is subject to lateral (transversal) as well as longitudinal spread. This phenomenon specifies a three-dimensional muon stopping region. The phenomenological formula for the lateral and longitudinal spread

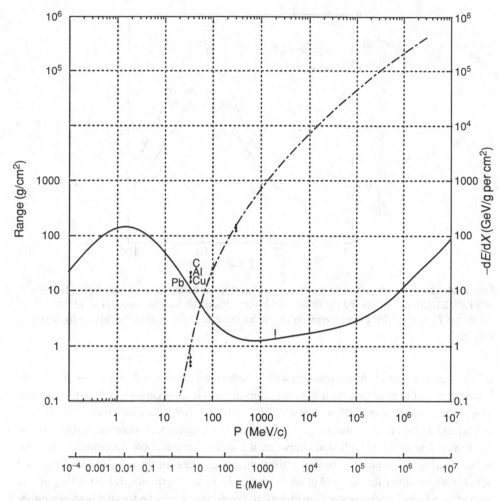

Figure 3.3 Mean range (broken line) and energy loss (continuous line) of the muon in lead (Pb), copper (Cu), aluminum (Al), and carbon (C) versus energy and momentum of the muon.

($D_{//}$ and $D_\perp$, both in cm) was derived as a function of range (R_0 in cm) (Fowler *et al.*, 1965), as $D_{//} = 2.6 \times 10^{-2} R_0^{0.94}$ and $D_\perp = 7 \times 10^{-2} R_0^{0.92}$. These semiempirical formulae can now be compared with the values obtained by the computer code GEANT, whose results are summarized in Figure 3.5.

Below energies of a few keV, a significant difference exists between the behavior of μ^+ and that of μ^-. For μ^+, depending upon the nature of the condensed matter in which it is stopping, there are two possibilities. In gases, insulators, and most semiconductors, at the end of the ionization track, the neutral bound state Mu (muonium, a hydrogen-like atom composed of μ^+ and e^-) is formed. Subsequently Mu is further decelerated via elastic collisions with the surrounding atoms. As seen in Figure 3.6, during the slowing-down process, there is a finite probability for the electron of Mu to be stripped and become

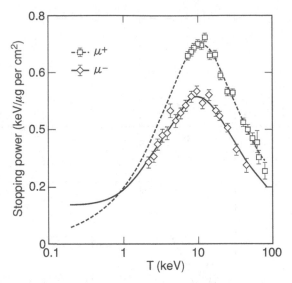

Figure 3.4 The energy loss (stopping power) of μ^+ and μ^- in carbon at an energy region below 100 keV. Experimental data and the fitting curve were obtained from Mühlbauer *et al.* (1999), where the T_∞ is specified as the energy region for an increase in the spectral density by frictional cooling.

μ^+ through a charge-exchange reaction with isolated inert gas atoms: $Mu + A \rightarrow \mu^+ + A^-$. It is interesting to note that, with the exception of He, electron capture dominates electron loss at the energy above a few 100 eV. This behavior holds in most forms of the soft condensed matter. In most metals, μ^+ thermalizes in a diamagnetic state; no stable Mu can be formed due to strong collisions between μ^+ and the conduction electrons. At the end of the slowing-down process, some equilibrium energy state for the μ^+ and Mu is reached which may be either thermalized at an energy of kT or nonthermalized. Depending on the type of condensed matter under consideration, bound states may be formed between either μ^+ or Mu and the atoms or molecules of the host material, as described in later sections.

The formation mechanism of Mu during the μ^+ slowing-down in condensed matter has been the subject of several experimental studies. As summarized in Figure 3.7, there are two extreme models – the hot atom model and the spur model. In the first of these, the Mu is formed at epithermal energies where the μ^+-e^- energy falls in the energy gap region known as the Ore gap, while in the second model the Mu is formed in a radiolysis process involving the interaction between thermalized μ^+ and an e^- generated (with a positive ion) during the slowing-down of the μ^+. In experiments conducted with an applied electric field (attracting or repelling electrons in the track against μ^+) in various insulators and semiconductors, excess electrons from the ionization track are mobile enough to reach the μ^+ and form Mu within the muon spin rotation/relaxation/resonance (μSR) time range (Krasnoperov *et al.*, 1992; Storchak *et al.*, 1997).

The degree of depolarization due to ionization during slowing-down can be estimated semiquantitatively using a formula given by Akylas and Vogel (1977). From this formula

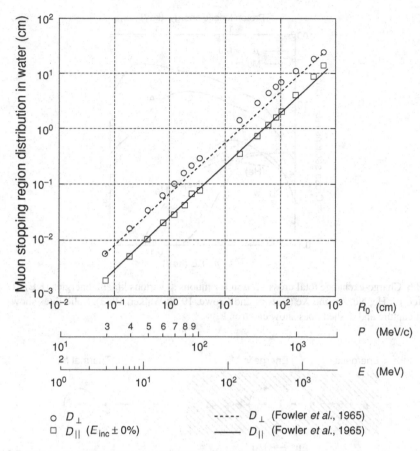

Figure 3.5 Lateral and longitudinal spread due to Coulomb scattering and straggling of a parallel pencil muon beam in water with energy E_0. Values estimated by the semiempirical formula (Fowler *et al.*, 1965) are compared with those (○, □) estimated by the computer code GEANT.

one can expect the spin precession occurring during the energy loss process from 1 MeV down to 1 keV for either μ^+ or μ^- to be less than 0.1 mstr, so that the depolarization is negligibly small.

The polarization of the μ^+ will change depending upon the characteristics of Mu formation at intermediate stages of the slowing-down process. Once the μ^+ is coupled with an unpolarized e^- from the condensed matter sample to form Mu, the polarization of Mu is immediately reduced to 50% as a result of the fact that only 50% of the μ^+ forms fully polarized triplet Mu with F (sum of spin angular momenta of μ^+ and e^-) = 1 while the other 50% of the μ^+ forms singlet Mu with $F = 0$. The final polarization is determined by the state in which μ^+ resides for a long time compared to a characteristic time corresponding to the Mu precession frequency ($\tau_{Mu} = \omega_{Mu}^{-1} = 0.4$ ns for Mu in vacuum). In most cases, the significant depolarization of Mu depends upon the fraction of the initial μ^+ which takes the form of Mu at the final thermalization process; depolarization of Mu, once it is formed, is not significant during the slowing-down process.

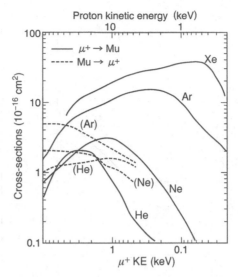

Figure 3.6 Charge-exchange total cross-section for muons at various kinetic energies (KE) as calculated for He, Ne, Ar, and Xe (Brewer and Crowe, 1978; Walker, 1983). Solid lines show electron capture and dashed ones show electron loss.

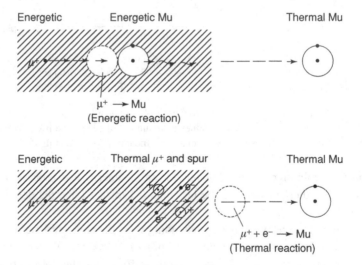

Figure 3.7 Summary of the formation mechanisms of Mu in various types of condensed matter.

The μ^-, on the other hand, reaching the end of the ionization regime, will be strongly attracted by the electric field from the nuclei. As described in Chapter 4, μ^- replaces an electron from the innermost atomic shell (the K shell) to form a muonic atom in an excited state with the critical quantum number $n_c \left(= \sqrt{m_\mu/m_e} = 14\right)$. Actual populations of each state of (n, ℓ_n) are widely distributed. There have been theoretical studies to predict these distributions, as described in detail in Chapter 4. Subsequently, by a process involving Auger transitions between the higher orbits and radiative transitions between the lower orbits, the μ^- cascades down to its ground state. As described later (Chapter 4), muonic hydrogen (μ^-p, μ^-d, and μ^-t) in high-density hydrogen is subject to other processes of cascade transitions.

Before the end of the ionization processes, the polarization of the μ^- follows a pattern similar to that of the μ^+ and remains nearly 100%. At the time of muonic atom formation, however, a significant change in the μ^- polarization occurs. In most cases, the final state of the atomic-capture process (the state of the formed muonic atom) is a state with a fine-structure splitting due to spin-orbit interactions, and consequently the μ^- capture process is spin-dependent. As a result, the μ^- polarization experiences a significant reduction. After forming a muonic atom in a state with angular momentum quantum numbers $J_\pm = \ell \pm \frac{1}{2}$, the polarization can be written as (Mann and Rose, 1961):

$$P_\mu(J_+) = \frac{1}{3}\left(1 + \frac{1}{J_+}\right)$$

$$P_\mu(J_-) = -\frac{1}{3}$$

Then, during cascades, depending upon the number of angular momentum quanta taken away by the transition radiation (1 for $E1$ (electric dipole), $M1$ (magnetic dipole), 2 for $E2$, $M2$ etc.), the polarization undergoes a change determined by a formula originally developed for use in experiments on oriented nuclei and perturbed angular correlation measurements (Nagamine and Yamazaki, 1974; Kuno et al., 1987).

For the low-lying states of a muonic atom formed with a nucleus of spin I, an interaction can occur between the muon spin and either the nuclear magnetic moment or the nuclear quadrupole moment, leading to a hyperfine splitting in level J of the muonic atom: $F = J \pm I$. The polarization of the F state then takes the following form:

$$P_\mu(F) = \frac{J}{2F+1}P_\mu(J)$$

Thus, at the end of the cascade process, the polarization of the μ^- in its 1s ground state takes a value roughly as follows.

$$P_\mu(1s_{\frac{1}{2}}) \approx \frac{1}{6} \qquad\qquad \text{for } I = 0$$

$$P_\mu(1s_{\frac{1}{2}}) \approx \frac{1}{6} \times \frac{1}{3}\left(1 + \frac{2}{2I+1}\right) \qquad \text{for } I \neq 0$$

With the use of polarized or oriented nuclei, the polarization of the muonic atom can be restored. The idea of repolarization was introduced by Nagamine and Yamazaki (1973); this was followed by experimental confirmation using polarized ^{209}Bi targets (Kadono et al., 1986), together with several detailed theoretical calculations assessing the relative importance of the various contributing processes (Kuno et al., 1987).

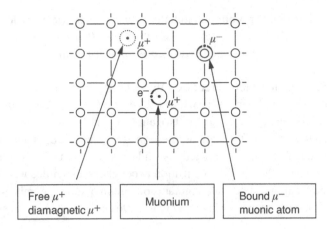

Figure 3.8 Schematic pictures of three typical states of muons in an atomic lattice after implantation into condensed matter.

3.2 Behavior of muons in matter

In its final state, at the end of the energy loss processes in condensed matter described in the previous section, the μ^+ or μ^- spends most of its lifetime in one of the forms depicted conceptually in Figure 3.8. The three states most frequently encountered in muon science studies are: diamagnetic μ^+, paramagnetic Mu, and muonic atom states of the μ^-. Most activities in muon science studies are concerned with phenomena relating to these long-lived final states of the μ^+ and μ^-. Some essential features of the properties of these states are given in the following section; more detailed descriptions will be found in later chapters as the relevant cases arise in each muon science topics.

3.2.1 Diamagnetic μ^+

In most metals, the μ^+ is in a diamagnetic state and resides at an interstitial site. Also, in other materials such as semiconductors or insulators, there can occur bonding states such as muonic oxide (μ^+-O^-) which behave like diamagnetic μ^+. The conduction electrons in metals screen the positively charged interstitial μ^+. However, there is no coherency in the screening electrons; there is no case in which a single electron is captured permanently into a screening orbit around the μ^+. There have been a series of thorough experiments to search for the existence of a paramagnetic Mu state in metals. So far, no positive results have been reported.

The nature of the electron screening around the μ^+ in metals can be investigated by measuring the paramagnetic Knight shift (K_p) of the μ^+ under applied field (H_{ext}). Without screening, the conduction electrons in a metal with paramagnetic susceptibility χ_p become uniformly polarized, contributing to the μ^+ internal magnetic field of a quantity $\frac{4\pi}{3}\chi_p H_{ext}$. With screening, there is an enhancement of the internal field by a factor

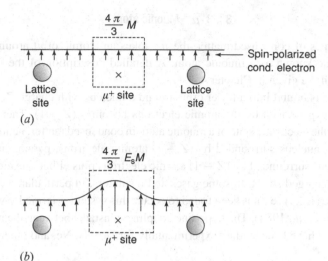

Figure 3.9 Schematic representations of screening enhancement parameters for the conduction electrons around the μ^+ in typical metals, as exhibited in Knight shift measurements. (a) No screening enhancement; (b) screening enhancement.

of $E_\mu = |\psi(0)|^2$; $K_p = (4\pi/3)E_\mu\chi_p H_{\text{ext}}$, yielding information on the type and degree of electron screening around the μ^+.

Systematic experimental studies by Schenck (1985) have explored the nature of conduction electron screening around the μ^+. A somewhat simplified summary of these studies is given in Figure 3.9.

3.2.2 Paramagnetic muonium

In insulators and most semiconductors, the stable state is a paramagnetic atom-like system, with a single electron in a bound orbit around the positively charged μ^+; this species, muonium, is exactly analogous to the ground state of the neutral hydrogen atom. Owing to the many-body effects present in matter, the effective mass or orbital radius of the paramagnetic electron can take a different value compared to that of muonium in vacuum. Revealing the detailed nature of these changes in the electronic structure of Mu in condensed matter as well as formation mechanisms of the Mu, as briefly described earlier, is the objective of various experimental and theoretical studies (see Chapters 6–8).

There are several possible charged states of the μ^+ in condensed matter, namely, $Mu^0(\mu^+ + e^-)$, $Mu^-(\mu^+ + 2e^-)$ and $Mu^+(\mu^+ + e^- + \text{hole})$. The existence of stable Mu^- has been confirmed in several semiconductors, while the hyperfine coupling constant of Mu has been studied experimentally in a wide range of semiconductors and insulators. As seen in Chapter 8, the radius of paramagnetic Mu depends upon the nature of the host material and scales with either the band-gap energy of the material or the material's ionicity.

3.2.3 μ^- Muonic atom

In almost all types of condensed matter, the μ^- takes an atomic orbit around a nucleus, yielding the state known as a muonic atom. A detailed description of the nature of the muonic atom will be given in Chapter 4.

When a muon is bound into a lower orbit around a nucleus with charge Z, the effective nuclear charge experienced by the atomic electrons becomes $(Z-1)$. There are several possibilities for the electronic state of a muonic atom in condensed matter: neutral atom with muonic $(Z-1)$ nucleus surrounded by $(Z-1)$ atomic electrons, paramagnetic $(Z-1)$ atom with nucleus surrounded by $(Z-1)$ atomic electrons plus either one electron or one hole, and other charged states. In some gases and insulators, in particular, a paramagnetic state of structure $(\mu^- Z)^+ e^-$ has been found to occur; this system is termed a Mu-nucleonic atom (Dobretzov et al., 1984). The hyperfine coupling constants between the muon and the unpaired electron have been studied experimentally for He, N_2, Ne, and other cases.

REFERENCES

Akylas, V. R. and Vogel, P. (1977). *Hyperfine Interact.*, **3**, 77.

Brewer, J. H. and Crowe, K. M. (1978). *Annu. Rev. Nuclear Particle Science*, **28**, 239.

Dobretzov, Yu. P. *et al.* (1984). *Hyperfine Interact.*, **17–19**, 845.

Fowler, P. H. *et al.* (1965). *Nature*, **189**, 524.

Kadono, R. *et al.* (1986). *Phys. Rev. Lett.*, **57**, 1847.

Krasnoperov E. P. *et al.* (1992). *Phys. Rev. Lett.*, **69**, 1560.

Kuno, Y. *et al.* (1987). *Nucl. Phys.*, **A475**, 615.

Mann, R. A. and Rose, M. E. (1961). *Phys. Rev.,* **121**, 293.

Mühlbauer, M. *et al.* (1999). *Hyperfine Interact.*, **119**, 305.

Nagamine, K. and Yamazaki, T. (1973). *TRIUMF-Proposal*, **E73**.

Nagamine, K. and Yamazaki, T. (1974). *Nucl. Phys.*, **A219**, 104.

Schenck, A. (1985). *Muon Spin Rotation Spectroscopy*. Bristol: Adam Hilger.

Storchak, V. G. *et al.* (1997). *Phys. Rev. Lett.*, **78**, 2835.

Walker, D. (1983). *Muon and Muonium Chemistry*. Cambridge: Cambridge University Press.

4
The muonic atom and its formation in matter

After the end of the slowing-down mechanisms described in Chapter 3, the μ^- takes the form of a muonic atom, entering an orbit around one of the atomic nuclei of the stopping material. Following the initial formation, in a cascade process within the atom, the μ^- reaches its atomic ground state and remains there until it decays into an electron or is captured by the nucleus. In this chapter the basic properties of the ground state of the muonic atom are described, and an outline is given of scientific research related to the formation mechanism of muonic atoms.

4.1 Basic properties of the ground state of muonic atoms

The ground state of a muonic atom, in which the μ^- passes most of its lifetime following its injection into materials, is characterized by the following properties: the binding energy, size, lifetime/nuclear capture rate, and magnetic moment.

When the μ^- forms a muonic atom with a light nucleus with atomic number Z, the radius $R_\mu(1s)$ and binding energy $E_\mu(1s)$ of the ground state can be described in the point nucleus approximation as follows:

$$R_\mu(1s) \cong 270/Z \times 10^{-13} \quad (cm)$$
$$E_\mu(1s) \cong 13.6 \times 207 \times Z^2 \quad (eV)$$

This remains a good approximation until the radius of the 1s orbital of the muonic atom becomes similar to that of the nucleus. The nuclear radius increases with increasing nuclear mass number A; $R_N \cong 1.2 \times \sqrt[3]{A} \times 10^{-13}$ (cm). Therefore, corrections become significant for nuclei heavier than Fe, Cu, etc. Several typical examples of R_μ and E_μ (1s) are summarized in Table 4.1a. More detailed descriptions can be found in several references (Engfer et $al.$, 1974; Hüfner et $al.$, 1977).

The energy levels of muonic atoms differ significantly from the predictions of the classical point-nucleus approximation as a result of the following two factors: (1) the finite-sized nature of the nuclear charge distribution; and (2) vacuum polarization. The first correction is obvious by comparing the values in Table 4.1a and $13.6 \times 207 \times Z^2$ (eV). Some examples of the vacuum polarization correction are given in Table 4.1b. Actually, the values of $E_\mu(nl)$ have long been used as a good measure of the nuclear charge distribution.

Table 4.1a Typical examples of the K_α transition energies and binding energy $E_\mu(1s)$ as well as the radius $R_\mu(1s)$ of the ground states of muonic atoms with the radius of core-nucleus r_N

Z	Nucleus	$E_\mu(2p-1s)$ [a] (keV)		$E_\mu (1s)$ [b] (keV)	$R_\mu (1s)$ [c] (fm)	$(r_N^2)^{1/2}$ [a] (fm)
		$2p_{3/2}-1s_{1/2}$	$2p_{1/2}-1s_{1/2}$			
2	^{4}He	8.228(4)		10.96	197.3	
4	^{9}Be	33.39(5)		44.51	97.22	2.62(91)
6	^{12}C	75.248(15)		100.37	64.68	2.49(5)
8	^{16}O	133.525(15)		178.30	48.53	2.71(2)
13	^{27}Al	346.82(15)		465.71	30.09	3.025(23)
15	^{31}P	456.54(20)		614.88	26.23	3.188(18)
26	^{56}Fe	1257.15(6)	1252.95(6)	1731.57	15.84	3.7143(57)
50	^{120}Sn	3464.78(47)	3419.06(40)	5220.33	9.662	4.642(6)
82	^{208}Pb	5963.77(45)	5778.93(50)	10596.51	7.431	5.4978(30)
83	^{209}Bi	6032.2(50)	5839.7(55)	10774.52	7.388	5.513(7)

[a] Engfer, R. *et al.* (1974). *Atomic/Nuclear Data Tables*, 14, 509.
[b] Barrett, R. C. (1977). *Muon Physics*, vol. I, ed. V. W. Hughes and C. S. Wu, p. 309. New York: Academic Press.
[c] Koike, T. (2001). Private communication.

Table 4.1b Typical examples of vacuum polarization correction $(E_{\mu(1s)}^{VP})$ in $E_{\mu(1s)}$ [a]

Z	Nuclear	$E_\mu (1s)$ (keV)	$E_\mu^{VP} (1s)$ (keV)
2	^{4}He	10.96	0.02
8	^{16}O	178.30	0.83
26	^{56}Fe	1731.57	11.83
82	^{208}Pb	10596.51	67.15

[a] Barrett, R. C. (1977). *Muon Physics*, vol. I, ed. V. W. Hughes and C. S. Wu, p. 309. New York: Academic Press.

Additionally, hyperfine structure in the energy levels of muonic atoms is a measure of the distributions of the magnetic moment ($M1$) and the quadrupole moment ($E2$). Typical examples are summarized in Table 4.2.

The lifetime of the ground state (τ_N) depends upon the Z number of the nucleus where there are two competing processes, free muon decay (Λ_d) and nuclear muon capture (Λ_c); $\tau_N^{-1} = \Lambda_d + \Lambda_c$. The elementary nuclear capture process is $\mu^- + p \rightarrow n + \nu_\mu$. Since the capture rate is proportional to the μ^- spatial density at the nucleus $[1/R_\mu(1s)]^3$

Table 4.2 Typical examples of the magnetic hyperfine splitting of the ground states and the first excited states of muonic atoms and the charge deformation parameter (β) for change distribution parameter (c, t) obtained from the electric quadrupole hyperfine interaction in 2p−1s and 3d−2p muonic transitions (Engfer *et al.*, 1974)

Nucleus	$A_1^{\exp}(1s_{1/2})^a$	$A_1^{\exp}(2p_{1/2})^b$	μ/μ_{NB}
^{93}Nb(9/2$^+$)	1.560(48)	0.374(20)	6.167
^{115}In(9/2$^+$)	1.61(13)	0.383(85)	5.5351
^{127}I(5/2$^+$)	0.89(9)	0.294(89)	2.8091
^{205}Tl(1/2$^+$)	0.61(3)	0.53(6)	1.62754
^{209}Bi(9/2$^+$)	2.16(15)	1.50(20)	4.0802

$^a A_1(I; 1; 0; 1/2) = \Delta E_{mag}(\exp)/(2I+1)/I$
$^b A_1(I; 2; 1; 1/2) = \Delta E_{mag}(\exp)/(2I+1)/I$

Z	Nucleus	c (fm)	t (fm)	β	$(r^2)^{1/2}$ (fm)	Q_0 (barn)
60	^{150}Nd	5.871(27)	2.342(57)	0.278(3)	5.048	5.15(10)
62	^{152}Sm	5.902(27)	2.364(53)	0.296(3)	5.090	5.78(10)
66	^{162}Dy	6.007(27)	2.404(53)	0.338(3)	5.211	7.36(10)
	^{164}Dy	6.109(33)	2.193(57)	0.334(5)	5.218	7.42(10)
67	^{165}Ho	6.27(10)	1.50(40)	0.30(1)		7.9(5)

(proportional to Z^3) as well as to the proton number of the nucleus (proportional to Z), Λ_c is proportional to Z^4 overall (the Z^4 law). Thus, the nuclear capture rate is given by the expression:

$$\Lambda_c = \Lambda_1 Z^4$$

where Λ_1 is the capture rate in the case of hydrogen. For heavier nuclei, however, the capture rate begins to differ significantly from the Z^4 law and Λ_c has a saturation value around 0.1 μs^{-1} at $Z \simeq 100$. For higher Z elements, there are significant corrections to be considered for the Z^4 law: (1) the radius of the first orbit $R_\mu(1s)$ is not proportional to Z^{-1} but subject to corrections due to the finite size of the nucleus; (2) the process of μ^- capture by a proton inside the nucleus can only allow the resultant neutron to occupy a previously unoccupied phase space region in the final state of the nucleus. Various theoretical and experimental studies have been carried out to obtain further correction terms to the Z^4 law (Primakoff, 1959; Goulard and Primakoff, 1974; Suzuki *et al.*, 1987). Typical examples are summarized in Table 4.3.

Because of the unique dependence of τ_N on Z, one can determine the Z value of the nucleus to which μ^- is bound by measuring the muon lifetime using the decay electrons. Thus, a type of elemental analysis can be performed by measurement of the decay-electron

Table 4.3 Models and conceptual approaches for the rate of μ^- capture via weak interactions in the ground states of muonic atoms

Formula	Parameters (Suzuki et al., 1987)
Primakoff	
$$\Lambda_c(A, Z) = Z_{eff}^4 X_1 \left[1 - X_2 \left(\frac{A - Z}{2A} \right) \right]$$	Z_{eff}: Ford, K.W. and Wills, J.G. (1968). *Nucl. Phys.*, **35**, 295. X_1: 170/s[1], hydrogen capture X_2: 3.125, Pauli exclusion effect
Goulard and Primakoff	
$$\Lambda_c(A, Z) = Z_{eff}^4 G_1 \left[1 + G_2 \frac{A}{2Z} - G_3 \frac{A - 28}{2Z} \right.$$ $$\left. - G_4 \left(\frac{A - Z}{2A} + \frac{A - 2Z}{8AZ} \right) \right]$$	G_1: 261 (252) G_2: -0.040 (-0.038) G_3: -0.26 (-0.24) G_4: 3.24 (3.23)

lifetime spectra. This elemental analysis capability is useful not only in the dedicated nondestructive elemental analysis technique to be described later in this chapter, but also more generally in the application of polarized negative muons to condensed-matter studies (the μ^-SR method), as described in Chapter 6.

The free decay rate Λ_d of the μ^- bound to the ground states of a muonic atom is subject to relativistic correction; mass correction due to binding energy causes a change in a free muon decay rate ($\Lambda_d^{free} = (G_F^2 m_\mu^5)/(192\pi^3)$, G_F: Fermi coupling constant of weak interaction). In heavy nuclei, this correction becomes large.

Free muon decay at the bound ground state, $\mu^- \rightarrow e^- + \bar{\nu}_e + \nu_\mu$, is subject to the additional corrections due to a limited phase space available for the emitted e^- to take. Thus, the e^- energy spectrum from bound μ^- is different from that from free μ^-; a suppressed distortion takes place at the high-energy region ($\sim$ (muon mass)/2), and the distortion becomes more significant for higher Z-nuclei (Porter *et al.*, 1951; Watanabe *et al.*, 1987). It was also theoretically predicted that the asymmetry of the emitted e^- from the polarized μ^- at the ground state of the muonic atom becomes accordingly different from that of the free μ^- decay (Gilinsky *et al.*, 1960; Watanabe *et al.*, 1987).

The magnetic moment of the ground state is subject to a correction due to relativistic motion of the μ^- around the nucleus. Characteristic values for this correction are given by the term introduced by Breit (1928), leading to the following estimate for a point nucleus of charge Ze: $(g_{free} - Z)/g_{free} \approx \frac{1}{3}(\alpha Z)^2 = \frac{1}{3}(v/c)^2$. More general expressions were given by Margenau (1940). Typical examples described in the reference literature (Yamazaki *et al.*, 1974) are summarized in Figure 4.1.

The nuclear muon capture process produces the emission of various particles as well as excited states of the final nuclei. The elementary primary process of $\mu^- + [p] \rightarrow n + \nu_\mu$ is followed by the nuclear cascade process induced by the energetic n produced by the primary process. Thus, in general, various processes like $\mu^- + A_Z \rightarrow B_{Z'} + xn + yp + z\alpha + \cdots$ occur. Experimental data and phenomenological analysis are summarized in a review article by Singer (1974). Recently, the importance of the μ^- capture process has been

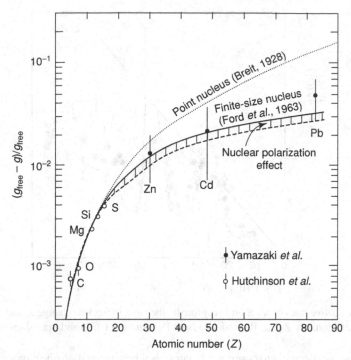

Figure 4.1 Magnetic moment of the μ^- bound in the ground state of common muonic atoms; experimental values and theoretical predictions. Experimental data are from Ford *et al.*, 1963; Hutchinson *et al.*, 1963; and Yamazaki *et al.*, 1973.

appreciated in cosmic-ray neutrino-related research as a need for information on radiation background due to the residual activities produced by nuclear-capture processes of the cosmic-ray muons at surrounding materials.

The capture reaction of the μ^- like ^{28}Si (μ^-, $2n$) can produce long-lived nuclei ^{26}Al ($t_{1/2} = 268$ y). Combined with the properties of cosmic-ray muons (see 2.5 and Chapter 10), one can use this reaction for the exposure dating of the earth-rock surface (containing SiO_2) against, e.g. ice coverage, with the help of accelerator mass spectroscopy to identify a tiny amount of ^{26}Al (Lal, 1991).

4.2 Muonic atom formation mechanism

When the μ is stopped in the target material made of a single element, as described in the previous chapter, the μ^-, after slowing down, takes a replacement with the innermost electron of the original element, forming the muonic atom at excited state with the critical quantum number n_{cri}; $n_{cri} \cong \sqrt{m_\mu/m_e} \cong 14$. Actual muonic states formed right after the replacement of the innermost electron take distribution of quantum numbers around n_{cri} and the associated orbital quantum number ℓ. Theoretical studies have been carried out on the population distributions of the states of the muonic atoms right after atomic capture, leading to relationships such as $P(\ell) \propto (2\ell + 1)$, etc. (Wightman, 1950; Baker, 1960; Haff *et al.*, 1974; Leon and Seki, 1977; Leon, 1980; Cohen, 1983, 1995, 1998, 2001).

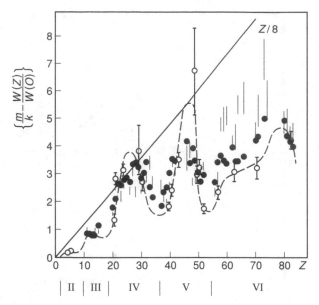

Figure 4.2 The relative rates of μ^- atomic capture by the atom of nuclear change Z and oxygen atom in oxides of stoichiometry $Z_k O_m$. Reproduced from Ponomarev (1973) and Stanislaus *et al.* (1987).

When the μ^- is injected into compound materials where more than one element is present, the population captured by each element depends on the capture process in a very specific way. The first model of the process of atomic capture of the μ^- in a molecule or solid-state system having different elements was put forward by Fermi and Teller (1947). Sometimes, the atomic capture phenomenon is referred to as the Fermi–Teller law.

The original Fermi–Teller law suggests that the populations $P(Z_1)$, $P(Z_2)$, $\cdots$ for molecules or other compound systems of stoichiometry $(Z_1)_m(Z_2)_n \cdots$ (e.g., for Fe_2O_3; $Z_1 = 26$, $m = 2$, $Z_2 = 8$ and $n = 3$) should have the following ratio:

$$P(Z_1)\colon P(Z_2)\colon \ldots\ldots = mZ_1\colon nZ_2\colon \ldots\ldots$$

Several experimental studies have been carried out to test the Fermi–Teller law. Systematic deviations from the simple law were noted. To take one example, in the case of oxides such as $Z_k O_m$, as seen in Figure 4.2, the relative population $(m/k)(P(Z)/P(O))$ is not $Z/8$, as the Fermi–Teller law predicts. Moreover, the deviation seems to follow the periodic table systematically (Stanislaus *et al.*, 1987).

Revised atomic capture laws have been proposed by Ponomarev (1973), Daniel (1975), Petrukhin and Suvorov (1976), Leon and Seki (1977), Schneuwly *et al.* (1978), and others. All of these theories have concentrated on refinements of the Fermi–Teller law. Daniel's theory has taken somewhat exact integration for an energy loss process of muon through Fermi gas. In the so-called "large mesic molecule" model by Ponomarev, redistribution of the muon initially occupying a molecular orbit of large radius before branching to populate each muonic atom state (inside this large "molecule") was considered and additional refinements have been made by Schneuwley *et al.*, taking into account redistribution possibilities of the μ^- at muonic molecular orbitals before muonic atomic orbit formation. In the "fuzzy

Table 4.4 Basic concepts of the various theoretical models for atomic capture by chemical compounds containing several elements

---

Model and formula
for μ^- capture in $Z_m Z_n^*$ in terms of capture per atom $R\,(Z/Z^*) = m^{-1}P(Z)/n^{-1}P(Z^*)$

Fermi–Teller
$$R(Z/Z^*) = Z/Z^*$$

Daniel
$$R(Z/Z^*) = Z^{1/3}\ell n(0.57Z)/Z^{*^{1/3}}\ell n(0.57Z^*)$$

Petrukhin–Suvorov
$$R(Z/Z^*) = (Z^{1/3} - 1)/(Z^{*^{1/3}} - 1)$$

Schneuley *et al.*
$$R(Z/Z^*) = P(Z)/P(Z^*), \quad P(Z) = \Sigma\rho(E_j)n_j(Z) + 2v\omega$$
where n_j is number of electrons in the jth sublevel of atom Z, $\rho(E_j)$ is the efficiency of their participation in the capture process, v is the valences of the Z atom and ω is the redistribution factor. For details, see Schneuwley *et al.* (1978).

Fermi–Teller" model by Leon and Seki, the orbital quantum number dependence for the μ^- atomic capture process is taken into account. The model by Petrukhin and Suvorov is based on experimental data for π^- in gas mixtures, and the proposed atomic capture rate is proportional to the squares of atomic radii ($Z^{1/3}$). The basic concepts of representative theories are summarized in Table 4.4. Further revisions were made by von Egidy *et al.* (1984) and Stanislaus *et al.* (1987).

4.3 Cascade transitions in muonic atoms

As seen in Figure 4.3, in the various cascade transitions occurring in the muonic atom (μ^-Z) after its formation by atomic capture of the muon in isolated atoms and molecules, there is a competition between radiative processes emitting X-ray photons $(\mu^-Z)_n \rightarrow (\mu^-Z)_{n'} + \gamma$ and (external) Auger processes which emit low-energy electrons from the atom's inner shell $(\mu^-Z)_n + Z' \rightarrow (\mu^-Z)_{n'} + Z'^+ + e^-$. Generally speaking, radiative transitions dominate over Auger transitions in cases where the transition energy gap is large (West, 1958). In addition, in some molecules, molecular dissociation $(\mu^-Z)_n + (Z')^2 \rightarrow (\mu^-Z)_{n'} + Z' + Z'$ becomes the dominant source of cascade transition (Borie and Leon, 1980). Also, Stark-mixing among different orbital quantum numbers must be taken into account: $(\mu^-Z)_{n,\ell} + Z' \rightarrow (\mu^-Z)_{n',\ell'} + Z'(n = n')$.

Some additional cascade transition processes are now known to be significant, in particular, in the case of hydrogen muonic atoms (μ^-p, μ^-d and μ^-t), which appear in the muon catalyzed fusion described in the following chapter. Coulomb deexcitation accelerates the muonic atom and elastic scattering decelerates the muonic atom. Coulomb deexcitation is $(\mu^-p, \mu^-d, \mu^-t)_i + (p, d, t) \rightarrow (\mu^-p, \mu^-d, \mu^-t)_f + (p, d, t)'(i > f)$ which becomes dominant over Auger transition for $n > 10$ (Bracci and Fiorentini, 1978; Ponomarev and Solov'ev, 1996). There, the muonic hydrogen has an energy gain of $\Delta E_{if}[m_p/(m_{\mu^-p} + m_p)]$ in the case of $(\mu^-p)_i + p \rightarrow (\mu^-p)_f + p'$. On the other hand, during the elastic scattering

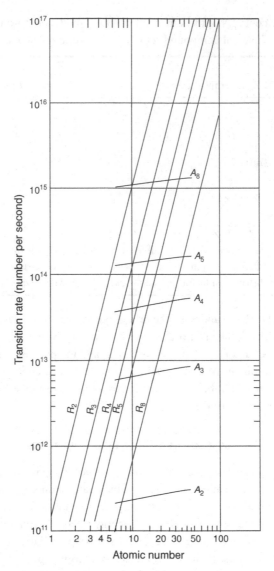

Figure 4.3 Rates of radiative transition (R) versus Auger transition (A) for various transitions in muonic atoms. Reproduced from West (1958).

process of $(\mu^- p, \mu^- d, \mu^- t)_i + (p, d, t) \rightarrow (\mu^- p, \mu^- d, \mu^- t)_f + (p, d, t)'$, due to a transport cross-section σ_n^t of $\pi (n^2 - 1)/m^* T$ at the state n with kinetic energy of T with reduced mass m^* (Menshikov and Ponomarev, 1988), muonic hydrogen has an energy loss $\Delta T/T$ of $\overline{(1 - \cos \theta)}(2m_{\mu^- p} m_H)/(m_{\mu^- p} + m_H)^2$ and an associated deceleration rate in the case of $(\mu^- p) + p$ elastic scattering at angle θ. As an additional acceleration mechanism of muonic/pionic hydrogen, the mechanism of muonic molecule formation via an Auger process followed by its predissociation was suggested (Menshikov, 1988; Kravtsov *et al.*, 2001).

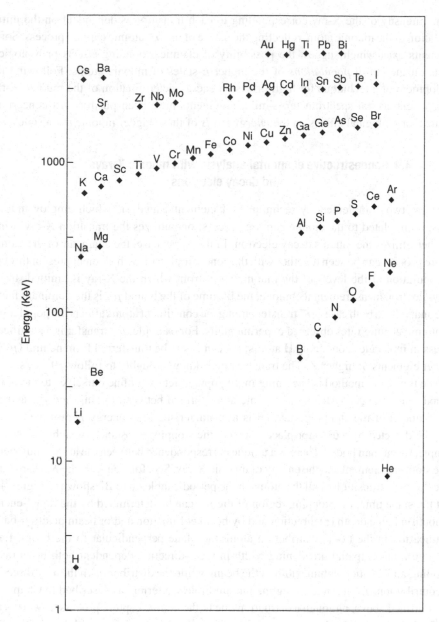

Figure 4.4 Typical transition energies of K_α X-ray in muonic atoms and relevant photon detectors.

The transition energy of cascade transitions for light nuclei can easily be estimated using the point-nucleus approximation. For heavier nuclei, as a result of the substantial corrections to $E_\mu(1s)$ and E_μ required to accommodate the effect of other low-lying levels, a more involved calculation is required. Some results for K_α X-ray of the 2p → 1s transition energy are summarized in Figure 4.4.

The intensity of the X-ray corresponding to each transition is dependent on the initial population of the muonic atom, reflecting the nature of the μ^- atomic capture process. Some arguments exist which suggest the possibility of chemical bonding effects or biological effects on the *initial populations* of the higher n-states of muonic atoms. Following the development of a technique for high-precision energy determination of the muonic X-ray using a bent-crystal spectrometer, similar arguments have been put forward concerning chemical or biological effects on the *energy levels* of these higher muonic atom states.

4.4 Nondestructive elemental analysis with muonic X-rays and decay electrons

There are two complementary techniques of elemental analysis which employ muonic atoms, both related to the atomic capture process; one utilizes the transition X-ray, while the other utilizes the muon's decay electron. In the X-ray case, the character of the atomic capture process can be seen together with the concentration of each element present through the population of the levels of the muonic atom from which the X-ray is emitted. In the decay-electron measurement technique, the lifetime of the bound μ^- is the dominant factor in the analysis. In both cases, μ^- transfer among the constituent atoms affects the populations and atomic capture rates observed experimentally. For example, μ^- transfer is a significant process in molecules containing H atoms; μ^- can easily be transferred from neutral (μ^-H) to other elements of higher Z. The transfer rate is known roughly to follow $10^{10}Z/\text{s}$.

Since the X-ray method is becoming more popular, let us confine ourselves to the X-ray method only. The X-ray from the muonic atom formed between the injected μ^- and the atomic nuclei of the stopping material is a characteristic high-energy photon which can easily be detected by a detector placed outside the stopping substance (which may be, for example, the human body). There is a unique correspondence between the atomic number Z of the stopping element and the energy of muonic X-ray. See, for example, the K_α X-ray line in the 2p $\rightarrow$ 1s transition of all the atoms in the periodic table up to Bi, shown in Figure 4.4.

At the same time, the stopping region of the μ^- can be determined by the range-energy distribution in the depth (z) direction and by beam collimation and/or beam position-based identification in the (x–y) distribution along the plane perpendicular to the beam. Each distribution has a spatial broadening width in the z-direction dependent both upon range straggling and the momentum width of the beam, while the distribution in the x–y direction has contributions from range straggling and multiple scattering, as described in Chapter 3.

As outlined above, although a correction due to the atomic capture process as well as the transfer from H has to be done, the rate of muonic atom formation is roughly proportional to the product of the Z value of the element and the elemental concentration $C(Z)$. Therefore, from the intensity of the muonic X-ray after correction for the energy-dependent efficiency one can obtain the elemental concentrations $C(Z)$ in the local region where the μ^- stops with at least qualitative accuracy.

Experimentally for the X-ray method, the selection of the relevant detector is essential. Depending upon the transition energies in muonic atoms, there are relevant detection methods for the transition photons. The significant feature is that most of the K X-ray in muonic

Table 4.5 Distribution of elements in the human body

Element	Abundance in human body (%)	Lifetime (μs)	X-ray energy (keV) (2p–1s)	X-ray energy (keV) (3p–1s)
C	23.0	2.026(2)	75.5	13.9
N	2.6	1.907(3)	102.0	18.9
O	61.0	1.795(2)	133.0	24.0
P	1.1	0.611(1)	458.0	88.4
Ca	1.4	0.333(3)	786.0	158.0

atoms can be detected by the most popular semiconductor detectors. Elemental analysis by the muonic X-ray method can be carried out using an experimental set-up of the type shown schematically in Figure 4.5, in which a beam collimator and a range-adjusting energy-absorber are placed selectively to control the stopping region of the μ^-.

Compared to the other elemental analysis methods, such as proton-induced X-ray element analysis (PIXE), secondary ion mass spectrometry (SIMS) or the neutron activation method, there are several advantages in the muon X-ray method. Since energies of the X-ray from low-lying transitions in the muonic atoms are significantly higher compared to the corresponding ones in the electronic atoms, the X-ray can easily be detected by the counters placed outside the objective materials to be analyzed; a nondestructive elemental analysis can be done for the portion which is situated deep inside the objective materials. This feature has a significant advantage over methods like PIXE or SIMS. One injected μ^- can produce at least one X-ray signal, while the induced radioactivities after the μ^- capture are relatively small. This feature has a strong advantage over the neutron activation method which usually produces a significant amount of residual radioactivity. Thus, a significant feature of the muon X-ray method can be summarized as the capability of the element analysis for the inside portion of precious material in a nondestructive way.

The human body is mostly composed of the elements listed in Table 4.5, where the corresponding energies of the muonic K_α and K_β lines are summarized. As a typical example of the potential use of muonic X-ray analysis in medical diagnostics, it has been applied to the case of osteoporosis, a disease known to be due to anomalous Al concentrations in the central trabecular part of the human backbone. Figure 4.6 shows the depth profile of the Al concentration in a "phantom" (model incorporating materials with appropriate densities and elemental compositions) of the human lumbar region, as measured in a Tohoku–Tokyo collaboration experiment conducted at Tri-University Meson Facility (TRIUMF) (Sakamoto et al., 1992; Hosoi et al., 1995).

The weak point of nondestructive elemental analysis with muonic X-rays, when applied to volumes located deep inside the object under study (e.g., some small organ within the human body), is the level of uncertainty as to the precise spatial region in which the μ^- stops. As described earlier (section 3.1), each distribution has a spatial broadening due to range straggling; additionally, the beam possesses a momentum width in the z-direction and shows broadening from both range straggling and multiple scattering contributions in

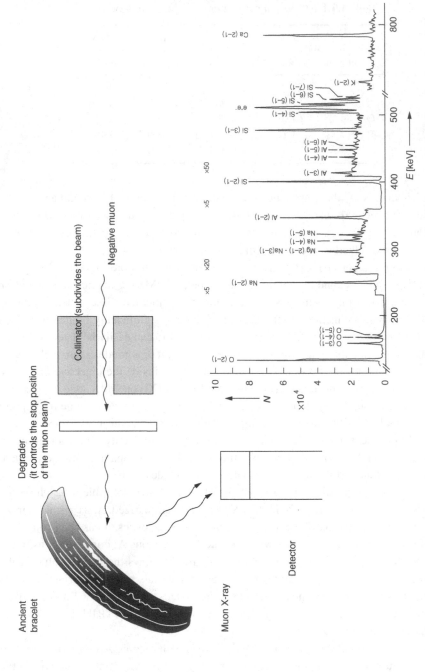

Figure 4.5 Experimental arrangement for the muonic X-ray method of elemental analysis for archaeological application (Daniel *et al.*, 1987).

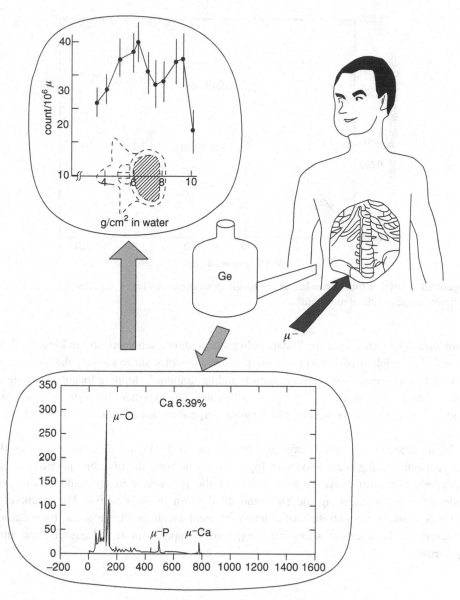

Figure 4.6 Schematic view of proposed diagnostic method for osteoporosis using the muonic X-ray technique, with experimental result on the corresponding phantom.

the x–y direction. A realistic case, for μ^- implanted into water with a momentum spread of 5%, is shown in Figure 4.7. There, for initial momentum 100 MeV/c corresponding to a range in water of 10 cm, both the longitudinal and the transverse spread become 3.0 mm. Recently, in order to overcome this difficulty a new advanced method of muonic X-ray elemental analysis was proposed by employing the Sudare collimator invented by M. Oda (private communication) and which has been applied to various astrophysical

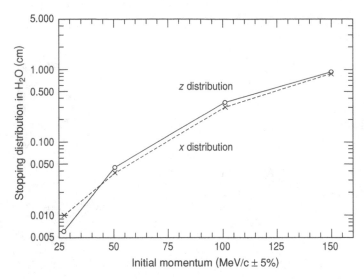

Figure 4.7 Spatial distribution in lateral and longitudinal directions for μ^- injected into water with different ranges of initial momentum.

problems. The principle of the Sudare collimator is shown schematically in Figure 4.8. Its essential principle is photon collimation through a detector slit to identify the narrow line connecting the photon source (here corresponding to the μ^- stopping location) to a given small detector region. Using this type of collimator, it is expected that a spatial resolution better than 0.1 mm can be achieved in the plane perpendicular to the line towards the photon detector.

Element analysis by the decay-electron lifetime method can be carried out using the experimental arrangement shown in Figure 4.9. The time distribution spectrum of the decay-electron with reference to the time of the μ^- arrival in the material can provide information regarding the elemental distribution in the material. For example, if there is a small concentration of a heavy element (such as Pb) in some hydrocarbon material, a characteristic short-lived component appears in the decay-electron time spectrum.

4.5 Future directions of muonic atom spectroscopy

As with future directions of X-ray spectroscopy of muonic atoms, several advanced experimental methods are under development.

4.5.1 Improving X-ray detection methods

In order to obtain more precise values of X-ray energies, several advanced methods have been proposed and realized. Distinguished examples are described below.

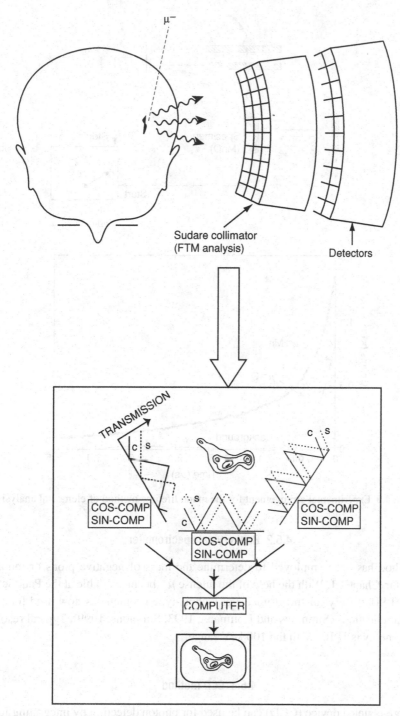

Figure 4.8 Experimental arrangement for future elemental analysis of the brain by measuring muonic X-rays using multiple Sudare collimators, with the principle of the Sudare collimator. FTM, Fourier transformation.

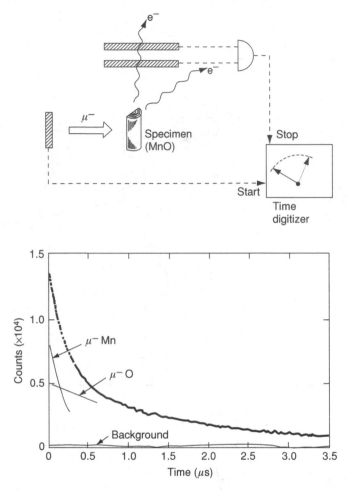

Figure 4.9 Experimental arrangement for the muon lifetime method of elemental analysis.

4.5.2 Bent crystal spectrometer

This method has been employed to determine the mass of negative pions or muons, as described in Chapter 1. With the help of the intense μ^- beam available at the Paul Scherrer Institute (PSI), the crystal spectrometer of Ochois-type geometry is now used for a wide range of applications (Simmons and Kottmann, 1993; Simmons, 1999). Typical resolution of X-ray energy is 1–10 eV in the 100 keV range.

4.5.3 CCD method

The charge-coupled device (CCD) can be used for photon detection by integrating ionized charges in detector materials quite accurately. The method was successfully applied to measure X-rays in the atomic process in muon-catalyzed fusion-related phenomena

(Egger, 1999). Typical energy resolution is 10–100 eV in the 10 keV energy range. As it takes $\sim$ 100 ms to integrate the charge, it is difficult to conduct a high-time resolution measurement.

4.5.4 Cryogenic calorimeter

Another type of high-energy-resolution photon detection is realized through a resistance change that occurs in some semiconductors at very low temperatures (e.g., below 0.1 K) due to X-ray energy deposition or by detecting superconducting tunneling current. This is called a cryogenic calorimeter (for recent developments, see Porter *et al.*, 2002), and typical energy resolution is 0.1 eV at a few keV range.

4.6 Exotic muonic atom systems

Application fields of muonic atom X-ray spectroscopy are now being extended in new directions. Several important examples are summarized here.

4.6.1 Muonic atom X-ray spectroscopy of unstable nuclei

X-ray spectroscopy of muonic atoms is an important tool to obtain information on nuclear charge distribution, since muonic X-ray transition energies are strongly affected by the size of nuclei. This method has been used successfully for many years to study stable isotopes in condensed or gaseous states. Recently, a new idea has been proposed (Strasser, 2001) to extend muonic atom X-ray spectroscopy to the use of nuclear beams, including radioactive beams, which would allow studies of the nuclear charge distribution of short-lived isotopes by means of the muonic X-ray method. Here, to combine a negative muon (μ^-) beam with an ion (Z^+) beam, and allow muonic Z atoms (μZ) to be formed, a common stopping medium made of a thin solid hydrogen (H_2 and/or D_2) film is used to stop simultaneously μ^- and Z^+, followed by the direct muon transfer reaction to higher Z nuclei to form a μZ atom. By adopting a double layer of mm thick solid H_2O with 0.1%D_2 and a few μm thick D_2, because of the Ramsauer–Townsend effect in (μd) + H collision, one can expect that injected MeV μ^- can be collected in a thin D_2 layer, where the collisional transfer will occur efficiently with the implanted Z-nuclei leading to longitudinal cooling.

REFERENCES

Baker, G.A. Jr. (1960). *Phys. Rev.*, **117**, 1130.
Borie, E. and Leon, M. (1980). *Phys. Rev.*, **A21**, 1460.
Bracci, L. and Fiorentini, G. (1978). *Nuovo Cimento*, **43A**, 649.
Breit, G. (1928). *Nature*, **122**, 649.
Cohen, J.S. (1983). *Phys. Rev.*, **27**, 167.
Cohen, J.S. (1995). *Phys. Rev.*, **A51**, 266.
Cohen, J.S. (1998). *Phys. Rev.*, **A57**, 4964.

Cohen, J.S. (2001). *Hyperfine Interactions*, **138**, 159.

Daniel, H. (1975). *Phys. Rev. Lett.*, **35**, 1649.

Daniel, H. *et al.* (1987). *Archaeometry* **29**, 1.

Egger, J.-P. (1999). *Hyperfine Interactions*, **119**, 291.

Engfer, R. *et al.* (1974), *Atomic/Nuclear Data Tables*, **14**, 509.

Fermi, E. and Teller, E. (1947). *Phys. Rev.*, **72**, 399.

Ford, K. W. *et al.* (1963). *Phys. Rev.*, **129**, 194.

Gilinsky, V. *et al.* (1960). *Phys. Rev.* **120**, 1450.

Goulard, B. and Primakoff, H. (1974). *Phys. Rev.*, **C10**, 2034.

Haff, P.K. *et al.* (1974). *Phys. Rev.*, **A10**, 1430.

Hosoi, Y. *et al.* (1995). *Br. J. Radiol.*, **68**, 1325.

Hüfner, J., Scheck, F., and Wu, C. S. (1977). In: *Muon Physics I*, ed. V. W. Hughes and C. S Wu, p. 202. New York: Academic Press.

Hutchinson, D. P. *et al.* (1963). *Phys. Rev.*, **131**, 1362.

Kravtsov *et al.* (2001). *Hyperfine Interaction*, **138**, 103.

Lal, D. (1991). *Earth Planet Sci. Lett.*, **104**, 424.

Leon, M. (1980). In: *Exotic Atoms 79*, ed. K. Crowe *et al.*, p. 141. New York: Plenum.

Leon, M. and Seki, R. (1977). *Nucl. Phys.*, **A282**, 445.

Margenau, H. (1940). *Phys. Rev.*, **57**, 383.

Menshikov, L. I. (1988). *Muon Catalyzed Fusion*, **2**, 173.

Menshikov, L. I. and Ponomarev, L. I. (1988). *Z. Phys.*, **D2**, 2.

Petrukhin, V. I. and Suvorov, V. M. (1976). *Zh. Eksp. Teor. Fiz.*, **70**, 1145.

Ponomarev, L. I. (1973). *Annu. Rev. Nucl. Sci.*, **23**, 395.

Ponomarev, L. I. and Solov'ev, E. A. (1996). *JETP Lett.*, **64**, 135.

Porter, C. E. and Primakoff, H. (1951). *Phys. Rev.*, **83**, 849.

Porter, C. E. *et al.* (1951). *Phys. Rev.*, **83**, 849.

Porter, F.S. *et al.* (2002). *AIP Conference Proc.*, **605**.

Primakoff, H. (1959). *Rev. Mod. Phys.*, **31**, 802.

Sakamoto, K., Hosoi, Y., and Nagamine, K. (1992). In: *Perspectives of Meson Science,* ed. T. Yamazaki, K. Nakai, and K. Nagamine, p. 487. North Amsterdam: Holland.

Schneuwly, H., Pokrovsky, V. I., and Ponomarev. L. I. (1978). *Nucl. Phys.*, **A312**, 419.

Simmons, L. M. (1999). *Hyperfine Interactions*, **119**, 281.

Simmons, L. M. and Kottmann, F. (1993). In: *Muonic Atoms and Molecules*, ed. L. A. Shaller and C. Petitjean, p. 307. Basel: Birkhäser.

Singer, P. (1974). *Springer Tracts*, **71**, 39.

Stanislaus, S. *et al.* (1987). *Nucl. Phys.*, **A475**, 642.

Strasser, P. (2001). *Nucl. Instr.*, **A460**, 451.

Suzuki, T., Measday, D. F., and Roalsvig, J. P. (1987). *Phys. Rev.*, **C35**, 2212.

von Egidy, T. *et al.* (1984). *Phys. Rev.*, **29A**, 455.

Watanabe, R. *et al.* (1987). *Prog. Theo. Phys.*, **78**, 219.

West, O. (1958). *Rep. Prog. Phys.*, **21**, 271.

Wightman, A. S. (1950). *Phys. Rev.*, **77**, 521.

Yamazaki, T. *et al.* (1973). *Phys. Lett.*, **53B**, 117.

5

Muon catalyzed fusion

5.1 Concept of muon catalysis of nuclear fusion

Of the two types of muons, only the μ^- is involved in muon catalyzed fusion (hereafter designated μCF) processes. As depicted in Figure 5.1, nuclear fusion reactions take place when two nuclei such as d and t approach one another to within the range of the nuclear interaction $r_n (\cong$ a few times 10^{-13} cm). However, because of the Coulomb repulsion between positively charged nuclei which increases with decreasing distance, the realization of nuclear fusion is not at all easy.

In the concept of thermal nuclear fusion, the additional energy is given by thermal energy (kT) through the satisfaction of the condition $kT \geq e^2/r_n$. By assuming $r_n \cong 10^{-12}$ cm, the right-hand side of the inequality becomes 7×10^4 eV (note that the radius and binding energy for the ground state of a hydrogen-like atom are 0.53×10^{-8} cm and 13.6 eV), the required temperature is 7×10^8 K (while room temperature, 300 K, corresponds to 0.03 eV). In the μCF concept, the fusion reaction is mediated by the neutral small atom formed between μ^- and a hydrogen isotope and the subsequent formation of a small muonic molecule, and the relevant energy is the appropriate overall formation energy.

Here, it might be relevant to mention significant features of fusion energy as a possible energy source in future centuries. They can be summarized as follows:

1. Compared to oil, coal, or natural gas, there are fewer problems in the fuel supply in a deuterium–tritium (D–T) fusion reactor since the deuterium (D) can be found almost unlimitedly in sea and river water (158 and 140 p.p.m., respectively) and the tritium (T) can be produced within a fusion energy process by using lithium (Li), which can also be found almost unlimitedly in the earth (8×10^8 t) and sea water (2.3×10^{11} t).
2. The exhaust of CO_2 gas is substantially smaller (less than 1%) so that it contributes to solve the earth's global warming problem.
3. Production of the radioactive product is significantly reduced compared to the conventional fission reactor.

In addition to possible future applications to energy resources the fascinating features in μCF research concern varieties of physical phenomena related to interplay between nuclear and electromagnetic interactions. These features are strikingly exemplified by the case of D–T mixtures with high-density ϕ comparable to the density of liquid hydrogen,

Table 5.1a Typical fusion reactions

p + d	→	^{3}He + γ	5.5 MeV
p + t	→	^{4}He + γ	19.8 MeV
d + d	→	^{3}He + n	3.3 MeV
		t + p	4 MeV
		^{4}He + γ	24 MeV
d + t	→	^{4}He + n	17.6 MeV
t + t	→	^{4}He + 2n	11.3 MeV
d + ^{3}He	→	^{4}He + p	18.4 MeV

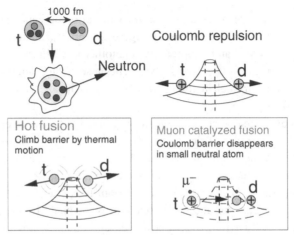

Figure 5.1 Conceptual view of the role of a negative muon used to remove the repulsive potential between d and t in order to catalyze nuclear fusion, with reference to thermal nuclear fusion.

$\phi_0(0.425 \times 10^{23}$ nuclei/cc). In the following, the present state of understanding of μCF and future prospects for μCF research are summarized, with particular emphasis being given to D–T μCF, and typical fusion reactions are listed in Table 5.1a.

The most basic phenomena of μCF consist of the following two processes: (1) the formation of a small muonic molecule and subsequent intramolecular fusion reaction; and (2) the mediation of a chain of fusion reactions by a single μ^-. These two processes are schematically summarized for the case of D–T μCF in Figures 5.1 and 5.2(a). Historically, to highlight the muon's catalytic role, the chain reaction has sometimes been presented in cyclical form, as shown in Figure 5.2(b). The basic process is summarized below. Details of the present level of understanding of each subprocess involved are given in later sections. After high-energy μ^- injection and stopping the μ^- in a D–T mixture, either a (dμ) or a (tμ) atom is formed, with a probability more or less proportional to the relative concentrations of D and T (C_d and C_t: $C_d + C_t = 1$). Because of the difference between (dμ) and (tμ) in the binding energies of their atomic states (either excited or ground), μ^- initially in the

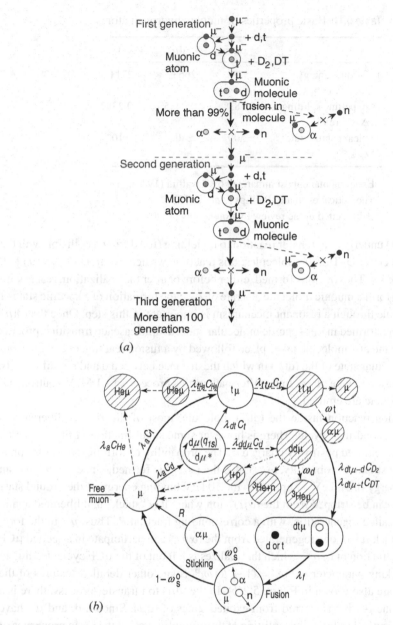

Figure 5.2 (*a*) The chain reaction of the muon catalyzed fusion phenomenon in a D/T mixture (above). (*b*) Cyclic representation of the chain reaction of the muon catalyzed fusion phenomenon in a D/T mixture, including the hyperfine effect and possible loss processes other than ($\alpha\mu$) sticking (below). Hatched region is the correction to the major μCF cycle.

Table 5.1b Basic properties of muonic hydrogen atoms

	pμ	dμ	tμ
1s binding energy E_{1s} (eV)	2528	2663	2711
1s hyperfine splitting ΔE_{hfs} (eV)	0.183[c]	0.0485[c]	0.2382[c]
Nuclear capture rate $\lambda_c(s^{-1})$	$\simeq$500[a]	$\simeq$400[a]	$\simeq$10[b]

[a] Experimental data summarized by Zavattini (1975).
[b] Theoretical estimate (Phillips *et al.*, 1975).
[c] Values cited in the review by Ponomarev (1990).

atomic state (dμ) undergoes a transfer reaction to t, yielding (tμ) during a collision with the surrounding t in either D–T or T_2 molecules; this reaction, written as $(d\mu) + t \rightarrow (t\mu) + d$, occurs at the rate λ_{dt}. The (tμ) thus formed, either before or after thermalization, reacts with D_2, DT, or T_2 to form a muonic molecule at a rate of $\lambda_{dt\mu}$; the formation of a specific state of the (dtμ) molecule through a resonant mechanism is important in this step. Once the (dtμ) molecule has been formed in this specific molecular state, a rapid cascade transition process of the μ^- inside the dtμ molecule takes place followed by a fusion reaction occurring from a low-lying molecular state of the (dtμ) in which the distance between d and t is sufficiently close to allow fusion to take place. In the aftermath of this process a 14.1 MeV neutron and a 3.5 MeV α-particle are emitted.

After the fusion reaction inside the (dtμ) molecule, most of the μ^- are liberated to participate in a second μCF cycle. There is however some small fraction of the μ^- which is captured by the emitted positively charged α. The probability of forming an $(\alpha\mu)^+$ ion is called the initial sticking probability, ω_s^0. Once the $(\alpha\mu)^+$ is formed, since the μ^- has an initial kinetic energy of 90 keV compared to the 10 keV binding energy of the ground state of $(\alpha\mu)$, the μ^- can be stripped from the $(\alpha\mu)^+$ ion where it is "stuck" and liberated again. This process is called regeneration, with a corresponding fraction R. Thus, μ^- in the form of either a nonstuck μ^- or one regenerated from the $(\alpha\mu)^+$ can participate in a second μCF cycle, while the fraction of $(\alpha\mu)^+$ which thermalizes is left out of the μCF cycle, leading to an effective sticking parameter ω_s: $\omega_s = (1 - R)\,\omega_s^0$. Some other detailed features of the dtμ-μCF cycle are also shown in Figure 5.2(b). In the (dμ) to t transfer process, there is a possibility that the μ^- is transferred from excited states of (dμ). Since t, d, and μ^- have spin, the muonic atom has hyperfine splitting at the ground state (see the basic properties of muonic hydrogen atoms summarized in Table 5.1(b). Therefore, there should be a hyperfine (spin-dependent) effect on the formation process of the muonic molecule. Also, the existence of a He impurity is inevitable, due to t-decay producing ^{3}He and the μCF process itself producing ^{4}He, and consequently μ^- loss due to capture by ^{3}He or ^{4}He must be taken into account.

Several different physical interactions are involved in the main processes of μCF. The fusion reaction taking place within the small muon molecule is the most significant part

Table 5.2 Major historical trends of muon catalyzed fusion (μCF) studies

1947 Hypothesis of the μCF cycle (Frank)
1948 Estimate of the fusion rate λ_f^{dd} (Sakharov)

1957 Observation of pdμ fusion (Alvarez *et al.*, 1957)
1957 Calculation of the dtμ cycle and sticking (Jackson)

1966 Observation of the T-dependence of $\lambda_{dd\mu}$ (Dzhelepov)
1967 Hypothesis of the resonant formation of ddμ (Vesman)

1977 Prediction of large $\lambda_{dt\mu}$ (Gerstein and Ponomarev)
1979 Observation of the upper limit on $\lambda_{dt\mu}$ and λ_{dt} (Dubna)
1979 Observation of the hyperfine effect in $\lambda_{dd\mu}$ (PSI)

1982 Measurement of $\lambda_{dt\mu}$, λ_{dt} (LAMPF)
1986 Observation of three-body effect in $\lambda_{dt\mu}$ (LAMPF, PSI)
1987 Observation of X-rays from $(\mu\alpha)^+$ in D-T μCF (PSI, KEK)
1987 Observation of X-rays from dHeμ (KEK)

1993 Observation of a large $\lambda_{dd\mu}$ in solid D_2 (TRIUMF)
1994 Observation of X-rays from muon transfer (PSI, KEK)
1995 Observation of $\lambda_{dt\mu}$ with eV(tμ) (TRIUMF)
1997 Systematic studies of X-rays, neutrons from D-T μCF (RIKEN-RAL)
2001 Measurements of $\lambda_{dt\mu}$ at high T and medium ϕ (JINR)
    Observation of anomalous T-dependence in $\lambda_{dt\mu}$ and W in solid D-T (RIKEN-RAL)

PSI, Paul Scherrer Institute; LAMPF, Los Alamos Meson Physics Facility; KEK, High Energy Accelerator Research Organization; TRIUMF, Tri-University Meson Facility; RIKEN-RAL, Institute of Physical and Chemical Research–Rutherford Appleton Laboratory branch; JINR, Joint Institute for Nuclear Research.

of the process where the nuclear interaction dominates, though there is also some nuclear interaction effect upon muon sticking and related processes. The remaining components of the cycle hinge mainly on electromagnetic interactions, and the basic role of the μ^- in these processes can be understood by considering it to be a heavy electron with a mass ratio m_μ/m_e of 207.

The concept of the μCF was introduced independently by Frank (1947) and Sakharov (1948). An experimental observation of P–D μCF was made by Alvarez *et al.* (1957) at Berkeley. The major historical trends in μCF studies are summarized in Table 5.2. Several review articles are available on the topic of μCF phenomena (Breunlich *et al.*, 1989; Ponomarev, 1990; Nagamine and Kamimura, 1998).

Let us summarize the basic important favorable features of μCF as a possible fusion energy source: (1) in common with other fusion energies such as thermonuclear fusion, inertial fusion, etc., the μCF is a clean energy source with fewer problems of fuel supply than nuclear fussion or fossil fuel use; (2) the μCF does not need high temperatures; (3) since it is initiated by the operation of the accelerator, the μCF is a fully controlled energy source without the phenomenon of criticality.

5.2 The experimental arrangements for muon catalyzed fusion

Once a reasonably intense beam of the μ^- becomes available, the experiments on the μCF phenomena can be realized with the following arrangements. Introduce the μ^- either as a single event of the continuous beam or as multiple events of the pulsed beam. The particles (fusion neutron, fusion proton, α-particle, etc.) or photons emitted during the fusion reaction can be detected with reference to the time of μ^- introduction. Usually, the electron from μ^- decay after the fusion cycle can be used either for the purpose of the muon number determination (event normalization) or for muon-related event identification.

Thus, the experimental set-up, as depicted in Figure 5.3, comprises the ending-part of the μ^- beam channel, target chamber, and detectors for the decay-electrons. Some details of the experimental arrangements for fusion products as well as the representative μCF experiments are described later as an aid to the understanding of each step of the μCF process.

The following remarks need to be made regarding detailed specifications of each detector, in particular, for fusion neutrons. With the occurrence of the μCF reaction inside the muon molecule, fusion neutrons are emitted with relatively high energy; e.g., 14.1 MeV in dtμ, $\sim$5 MeV in ttμ, and 2.5 MeV in ddμ. The standard neutron counter to be used is a liquid organic scintillator known under the commercial name of NE213. There, in order to eliminate the contribution of radiation background, the method of neutron-γ discrimination, based on the time-dependent characteristic of scintillation, must be employed. The energy calibration as well as efficiency calibration should be carried out using standard neutron sources available at the low-energy accelerator facilities. In the case of dtμ, because of the high-energy nature of the fusion neutron, detection is relatively easy, while in the case of ddμ, identification of the neutron signal against surrounding radiation background is not as easy, so that event identification with a coincidence signal from decay-electron is sometimes employed.

The use of a significant amount of radioactive tritium is inevitable in the case of D–T μCF studies. The 1 g of pure T_2 (1/6 mol, 3.6 ℓ STP and 10 cc in liquid) corresponds to 3.7×10^{14} Bq (10^4 Ci). Special precautions are needed when handling radioactive tritium. Usually, as realized at the Paul Scherrer Institute (PSI), the Institute of Physical and Chemical Research–Rutherford Appleton Laboratory branch (RIKEN-RAL) or the Joint Institute for Nuclear Research (JINR), a special T_2 gas-handling system is required for storage and transport of T_2-containing gas mixture. Also, since β-decay of t produces ^{3}He impurity at a rate of 100 p.p.m./day, it is important to install the ^{3}He removal system by employing a Pd-filter. Some descriptions of the updated tritium-handling system presently in use for the μCF experiment are available (Matsuzaki et al., 1999; Yukhimchuk et al., 1999).

5.3 Fusion reaction in a small muonic molecule

Now, let us consider what this small molecule (dtμ) looks like. The μ^- in the ground state of muonic hydrogen (μ^-p) is known to have an orbital radius of 260 fm and a binding energy of 2.5 keV. By analogy with the conversion from H(1s) to H_2^+ (g.s.), where the radius doubles

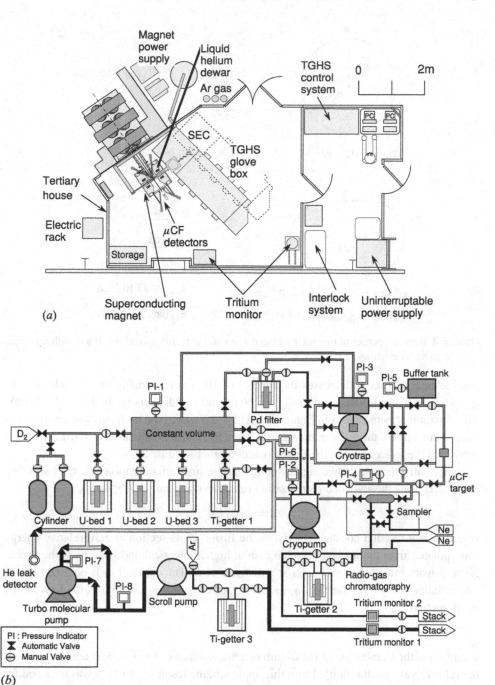

Figure 5.3 Typical examples of (a) the experimental arrangements for the muon catalyzed fusion (μCF) studies with pulsed μ^- beam at the Institute of Physical and Chemical Research–Rutherford Appleton Laboratory branch (RIKEN-RAL) and (b) diagram of the tritium gas-handling system (Matsuzaki et al., 2002). SEC, Secondary Enclosure Clean-up System; TGHS, tritium gas-handling system. Reproduced from T. Matsuzaki et al. (2001).

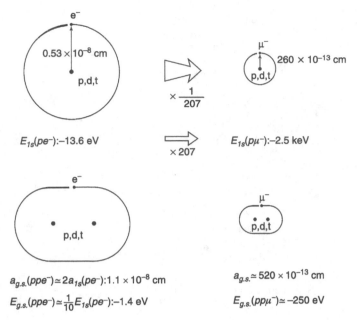

Figure 5.4 Basic properties of muonic hydrogen and muonic hydrogen molecular ion with reference to the equivalent electronic species.

and the binding energy decreases by a factor of 10, it is reasonable to conclude that the $(pp\mu^-)^+$ molecular ion has a radius of 2×260 fm and a binding energy of $2.5 \times (1/10)$ keV. The situation is summarized in Figure 5.4. Thus, the size of the molecule does not greatly exceed the range of the nuclear interaction (a few fm) and so, with the help of the zero-point motion of the molecular ion, the fusion reaction proceeds quickly.

Historically, the rate of nuclear fusion inside the small muonic molecule (λ_f) was first calculated using the so-called factorization relations (Jackson, 1957), that is:

$$\lambda_f = a_f |\psi(R)|^2$$

where a_f is a reaction constant related to the fusion cross-section at zero relative energy (interpolated from the nuclear reaction data at higher energies) and $|\psi(R)|^2$ is the probability density to find the two nuclei at a distance of R. The constant a_f is obtained by the interpolation ($v \to 0$) of the fusion reaction together with a description of the low-energy cross-section:

$$\sigma = a_f C_0^2 v^{-1}$$

where C_0 is the Gamow factor (Coulomb penetration factor) for s-wave scattering and v is the relative velocity at infinity. From this approach, the result $\lambda_f \sim 10^{12} \mathrm{s}^{-1}$ was obtained for D–T μCF.

There are several shortcomings in factorization treatments of the fusion rate. First of all, we need to know the fusion rate for the level of the muonic molecule specified by the rotational quantum number (J) and the vibrational quantum number (v) since, as described later, the muonic molecule is formed in an excited state. Second, distortion of the

Table 5.3a Theoretical fusion rates of muon catalyzed fusion at the levels of $(dt\mu)$ and $(dd\mu)$ molecules (s^{-1}) with level energies (eV)

Author	State (Jv)			
	(00)	(01)	(10)	(11)
$dt\mu$				
Bogdanova et al.[a]	1.0×10^{12}	0.80×10^{12}	1.1×10^{8}	4.2×10^{7}
Struensee et al.[b]	1.30×10^{12}	1.13×10^{12}		
Kamimura[c]	$(1.22-1.28) \times 10^{12}$	$(1.03-1.08) \times 10^{12}$	$(1.32-4.38) \times 10^{8}$	$(0.51-1.71) \times 10^{8}$
Szalewicz et al.[d]	1.25×10^{12}	1.05×10^{12}		
$dd\mu$				
Bogdanova [a]			4.3×10^{8}	1.5×10^{9}

[a] Bogdanova et al. (1988).
[b] Struensee et al. (1988).
[c] Kamimura (1989, private communication).
[d] Szalewicz et al. (1990/1991).

Molecule	(Jv)				
	(00)	(01)	(10)	(11)	(20)
$dd\mu$	325·074	35·844	226·682	1·97482	86·434
$dt\mu$	319·140	34·834	232·472	0·66017	101·416

molecular wave function due to nuclear interaction should be taken into account. Moreover, a further correction is needed in the formula for λ_f due to the dominance of a near-threshold resonance in the reaction cross-section.

Advanced calculations of the fusion rates at various levels of the muonic molecule $dt\mu$ were made by Bogdanova et al. (1988) and Kamimura (1989) using the complex nuclear potential (optical potential) method and by Struensee et al. (1988), Szalewicz et al., (1990/1991), and Hu et al. (1994) using the R-matrix method. All four types of calculations gave similar results concerning the fusion rates of the $J = 0$ states of $(dt\mu)$, as summarized in Table 5.3a. There, binding energies of (Jv) states of $(dt\mu)$ and $(dd\mu)$ are also presented. In comparison, fusion rates in the typical muon molecules other than $(dd\mu)$ and $(dt\mu)$ are summarized in Table 5.3b.

In order to understand the overall fusion rate in a muonic molecule, the detailed nature of the intramolecular cascade transitions must be known. As described later, the formation of muonic molecules occurs mostly via a resonant reaction which leads to an excited rotational–vibrational (Jv) state with $J = v = 1$ which is very weakly bound with respect to the $(t\mu)_{1s} + d$ threshold. The deexcitation of the muonic molecule proceeds via Auger transitions, according to:

$$[(dt\mu)_{Jv} \, d2e]^* \rightarrow [(dt\mu)_{J'v'} de]^* + e$$

Table 5.3b Fusion rates of muon catalyzed fusion at the levels of typical muon molecules (s^{-1}) other than (dtμ) and (ddμ) with level energies (eV)

Molecule (Jv)	Reaction channel	Ratio (%)	Theory (s^{-1})	Experiment
pdμ(00)			$8 \times 10^{5\,a}$	
	μ^3He + γ	86	$9.7(1) \times 10^{5\,b}$ $(\lambda_{t,\gamma}^{1/2})$	$3.5(2) \times 10^{5\,c}$
			$1.07(6) \times 10^{5\,b}$ $(\lambda_{t,\gamma}^{3/2})$	$1.1(1) \times 10^{5\,c}$
	^{3}He + μ	14	$0.62(2) \times 10^{5\,b}$ $(\lambda_{t,\mu}^{1/2})$	$0.56(6) \times 10^{5\,d}$
ptμ(00)	μ^4He + $\gamma(e^+e^-)$	95	$1.3 \times 10^{6\,a}$	$6.5(7) \times 10^{4\,e}$
	^{4}He + μ	5	1.3×10^5	
ttμ(11)	μ^4He + 2n	14	$1.2 \times 10^{7\,a}$	
(10)	^{4}He + 2n + μ	86	$1.3 \times 10^{7\,a}$	$1.5 \times 10^{7\,f}$
d^{3}Heμ				
(J = 0)			$10^{2\,g}$	
(J = 0)			$3(1) \times 10^{8\,h}$	
(J = 1)			$6(3) \times 10^{5\,h}$	

[a] Bogdanova (1982, Bogdanova et al., 1988), [b] Friar et al. (1991), [c] Petitjean et al. (1990/1991), [d] Bogdanova et al. (1990/1991), [e] Baumann et al. (1987), [f] Breunlich et al. (1987), [g] Kravtsov et al. (1984), [h] Kamimura, M. (private communication, 1989).

Molecule	(00)	(01)	(10)	(11)	(20)	(30)
ppμ	253·152	–	107·266	–	–	–
pdμ	221·549	–	97·498	–	–	–
ptμ	213·840	–	99·127	–	–	–
ttμ	362·910	83·771	289·142	45·296	172·652	48·813

The rates of these Auger deexcitation processes of the muonic molecule have been theoretically estimated (Bogdanova, 1982; Vinitsky et al., 1982). In Figure 5.5, the fusion reaction rates and the cascade transition rates for dtμ and ddμ molecules are summarized; in the dtμ molecule 80% of the fusion takes place from the (Jv) = (01) state and 20% from the (00) state, both of which are formed after cascading down from the (11) state. Combining all these arguments on the rates of fusion and deexcitation, we can conclude that the fusion reaction is completed in the muonic molecule in a time of 10^{-11} s (corresponding to a rate of 10^{11} s^{-1}) after the formation of the (11) state of the dtμ molecule during a collision between (tμ) and D$_2$.

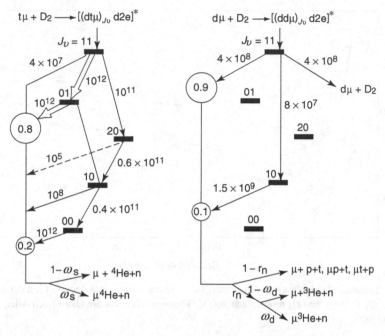

Figure 5.5 Scheme of cascade processes in dtμ and ddμ molecules after resonant molecular formation in the (1, 1) state, calculated by Bogdanova *et al.* (1982).

As is also shown in Figure 5.5, in the case of ddμ, which is also formed in the $(Jv) = (11)$ state, the fusion reaction takes place from the (11) and (10) states at a rate of around 10^8 s^{-1}. In this case of identical nuclei, molecular transition of $J = 1 \rightarrow J = 0$ is suppressed due to the Pauli principle preventing change of the nuclear spin, so that fusion in ddμ occurs from the p-wave of relative motion of nuclei. A similar situation takes place in ttμ fusion.

The branching ratio of the ddμ fusion, $R = N(^3\text{He} + \text{n})/N(\text{t} + \text{p})$ is known to be asymmetric 1.450(11) for $J = 1$ (Voropaev *et al.*, 2001). Theoretical understanding was given by Coulomb-corrected R-matrix calculation (Hale, 1990/91).

5.4 Neutral muonic atom thermalization

All the processes in μCF are initiated by the injection of μ^- into D–T (D$_2$ and T$_2$ mixture; chemically D$_2$, T$_2$, and DT) at high energies (MeV or higher). Thereupon, as described in Chapter 3, the μ^- slows down through ionization processes and is eventually captured by a d or t nucleus to form a muonic d or t atom. The neutral atoms thus formed are further slowed down through a series of elastic collisions with the surrounding medium and eventually move towards thermalization.

Competing processes involving neutral dμ and tμ can be quantitatively described in the form of energy-dependent cross-sections. Originally, in elastic scattering calculations involving dμ and tμ, it was assumed that the muonic atoms collide with bare d and t nuclei. However, the discrepancy between these calculations and experiments makes it clear that a realistic calculation is required to accommodate the condition that the nucleus

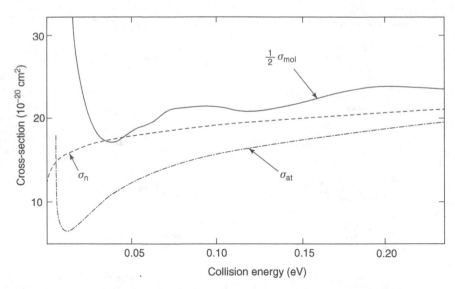

Figure 5.6 Theoretical calculations of low-energy elastic scattering cross-section for a $d\mu$ atom with d nucleus (σ_n), D atom, and D_2 molecule at 300 K. Theoretical curves are taken from Adamczak and Melezhik (1988).

of the collision partner is situated inside a D_2, DT, or T_2 molecule. In Figure 5.6, typical theoretical calculations (Adamczak and Melezhik, 1988) are summarized. Also, in actual μCF experiments, the hydrogen isotopes used are in condensed phases (liquid or solid). In experiments with D_2 in the solid phase, the thermalization mechanism proved to be different from that in the gas phase; as a result of the existence of an "energy gap," the thermalization of $(d\mu)$ in solid D_2 may not be complete.

The elastic scattering processes leading to thermalization compete with other processes, such as: (1) μ^- transfer to heavier nuclei such as from d to t or to higher Z impurities; and (2) muonic molecule formation. In cases where these processes occur, the thermalization process is interrupted, i.e., the muonic molecule is formed at epithermal energy.

The existence of nonthermalized muonic/pionic atoms of hydrogen isotopes has been confirmed in several experiments:

1. During the course of a precise measurement of the π-μ mass difference from the decay of π^- in liquid H_2, the $(p\pi)$ atom was found to have an epithermal energy (Crawford *et al.*, 1991).
2. As described later, in solid D_2 μCF the experimental formation rate of the ddμ muonic molecule was found to deviate markedly from theoretical predictions based upon resonant molecular formation; the result can be qualitatively accounted for by allowing for the presence of nonthermalized $(d\mu)$.

It is possible to make a direct measurement of the cross-section for elastic scattering of $(d\mu)$ + d using energetic $(d\mu)$ available from the Ramsauer–Townsend effect. Suppose that μ^- is introduced into a solid layer of H_2 containing a small amount of D_2 impurity (around 0.1%). After μ^- stops and forms $(p\mu)$ atoms, there is some probability of a μ^-

transfer reaction from $(p\mu)$ to d, producing energetic $d\mu$ due to the difference in binding energy. In experimental studies the energetic $(d\mu)$ was found to be emitted from the surface of the $H_2 + D_2$ layer. These energetic $(d\mu)$ can be used to study the elastic scattering process (Marshall $et\ al.$, 1990). If a second thin D_2 layer is deposited on to the $H_2 + D_2$ layer, since the $(dd\mu)$ fusion takes place at thermalized energies one can learn, by detection of a fusion product such as, e.g., the 3 MeV proton, the range of $(d\mu)$ as a function of the D_2 thickness, and this can in turn be converted into the energy dependence of the elastic scattering of $(d\mu) + d$ in solid D_2. The experimental result obtained demonstrates the importance of molecular and/or condensed matter effects (Strasser $et\ al.$, 1996).

During the cascade transitions in the muonic hydrogen (p, d, t), there is a mechanism for the muonic atom to be accelerated, as described in Chapter 4. This acceleration mechanism is now thought to be one of the reasons for the existence of energetic reactions (Ponomarev $et\ al.$, 1996).

5.5 Muon transfer among hydrogen isotopes

In a D–T mixture with a density ϕ of around ϕ_0, after injection at MeV energies, it takes 10^{-10} s for the μ^- to reach its ground state of either $(d\mu)$ or $(t\mu)$. After that, the μ^- remains in the ground state for most of its lifetime, as the nuclear capture rate to either d or t is negligibly small (400 s^{-1} for $d\mu$). Since the ground-state energy of $(t\mu)$ is deeper than that of $(d\mu)$ by 48 eV, a μ^- initially in a $(d\mu)$ atom in its ground state can easily be transferred through the reaction $(d\mu) + t \rightarrow (t\mu) + d$ via a collision with t in either T_2 or DT. Since the transfer rate of the μ^- is comparable to the radiative transition rate in $(d\mu)$ or $(t\mu)$, there is a possibility for the μ^- to undergo a transfer reaction while in an excited state. The probability for the μ^- to reach the ground state before transfer is denoted q_{1s}, and the problem of μ^- excited-state transfer is sometimes referred to as the q_{1s} problem (Ponomarev, 1983); $q_{1s} = 1$ corresponds to μ^- transfer after the μ^- has reached the ground state. Moreover, it has been pointed out that the $(d\mu) \rightarrow (t\mu)$ transfer reaction might occur at an epithermal energy with respect to $(d\mu)$. Some qualitative arguments relating to the q_{1s} values at various C_T, ϕ, and $E(d\mu)$ are summarized in Figure 5.7.

Experimentally, it is possible to measure the value of q_{1s} directly using the difference in the energy of, e.g., the K_α X-ray either between $(d\mu)$ and $(t\mu)$ for $(d\mu)$ to t transfer or between $(p\mu)$ and $(d\mu)$ for $(p\mu)$ to d transfer. Depending upon which X-ray is detected, one can simply conclude whether the transfer occurs from excited states or from the ground state. Such an experiment has become feasible with the development of high-resolution X-ray spectrometers. So far two experiments have been carried out for the $(p\mu)$ to d transfer: (1) using charge-coupled devices (CCD) at PSI (Lauss $et\ al.$, 1996) and (2) using seven-channel segmented small Si(Li) detectors at High Energy Accelerator Research Organization–Meson Science Laboratory (KEK-MSL) and at RIKEN-RAL (Sakamoto $et\ al.$, 1996). The results have shown the importance of existence of energetic $(p\mu)$ for its role in the transfer to d from excited states of $(p\mu)$.

Theoretical studies on the transfer reaction among hydrogen isotopes have been carried out and extended to cover the q_{1s} problem, the energy dependence of the initial state, and

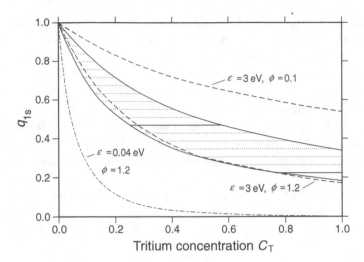

Figure 5.7 Theoretical prediction of the q_{1s} values in the $(d\mu) + t \rightarrow (t\mu) + d$ transfer reaction (Czaplinski *et al.*, 1994) together with the experimental values (shaded region) extracted from neutron data in $dt\mu$-μCF as a function of C_t at various $(d\mu)$ energies and densities (Ackerbauer *et al.*, 1996).

other details. The general tendency is for the experimental values of q_{1s} in D–T to be systematically larger than the theoretical values. This can be accounted for by considering the possibility of the existence of side-paths; the excited $(t\mu)$ states formed via transfer reactions from the excited $(d\mu)$ states collide with D_2 and resonantly form $(dt\mu)^*$ molecules which mostly decay into $(t\mu)_{g.s.} + d$. The net result is the apparent formation of additional $(t\mu)_{g.s.}$ (Froelich *et al.*, 1995).

5.6 Formation of muonic molecules

The small neutral $(t\mu)$ atom may closely approach a d nucleus in D_2 or DT, leading to the formation of a $(dt\mu)^+$ molecular ion, the ground state of which is, as described in section 5.1, small in size ($2a_\mu = 520$ fm) and tightly bound ($0.1E_\mu \approx -300$ eV).

Usually, the rate of formation of a tightly bound molecular state is relatively low. The most promising way is so-called Auger capture, which proceeds according to:

$$(d\mu) + D_2 \rightarrow [(dd\mu)de]^+ + e^-$$
$$(t\mu) + D_2 \rightarrow [(dt\mu)de]^+ + e^-$$

The theoretically predicted rate where the final state is either $(dd\mu)_{g.s}$ or $(dt\mu)_{g.s}$ is fairly low, of the order of 10^6 s^{-1} (comparable to the μ^- free-decay rate).

However, Gerstein and Ponomarev (1977) and Vinitsky *et al.* (1979) have theoretically predicted that an extremely shallow bound state with both the rotational and the vibrational quantum numbers equal to one (i.e., $(Jv) = (11)$) exists at an energy of $\varepsilon_{11} \approx -0.6$ eV, measured from the threshold energy of $(t\mu)_{1s} + d$. Because of the existence of this shallow bound state, substantially enhanced formation rates are expected through the following reaction process (known as resonant molecular formation), as can be seen in Figure 5.8:

$$(t\mu) + D_2 \rightarrow [(dt\mu)_{11} \, d2e]^* \qquad (\lambda_{dt\mu-d})$$

$$(t\mu) + DT \rightarrow [dt\mu]_{11} \, t2e)]^* \qquad (\lambda_{dt\mu-t})$$

Experimentally, the formation rate of the muonic molecule can be obtained through the relations between the observed overall cycling rate λ_c and the rates of the individual processes, e.g., λ_{dt} and $\lambda_{dt\mu}$, as shown diagrammatically in Figure 5.2(b). Let us consider μCF taking place in a D–T mixture high in both density and in C_t, and further assume that the atomic capture rates ($\lambda_{d\mu}$ and $\lambda_{t\mu}$) and fusion rate (λ_f) are high compared to the muon decay rate ($\lambda_{d\mu}, \lambda_{t\mu}, \lambda_f \gg \lambda_0$). The inverse of the cycling rate (λ_c^{-1}) then corresponds to the waiting time of $d\mu$ for muon transfer to t together with the time taken to form the molecule, i.e.,

$$\frac{1}{\lambda_c} \cong \frac{q_{1s} C_d}{\lambda_{dt} C_t} + \frac{1}{\lambda_{dt\mu} C_d}$$

Here, the factor $q_{1s} C_d$ is the probability that the muon reaches the ground state of $d\mu$, reflecting the fact that the transfer rates from the excited states of $(d\mu)$ are very rapid. In most of the D–T μCF experiment, as described elsewhere (Ishida et al., 2003), q_{1s} is assumed to be parameterized as $(1 + a_q C_t)^{-1}$ and $\lambda_{dt\mu}$ is taken to be $\lambda_{dt\mu}^0$ where $\lambda_{dt\mu}^1$ is neglected based upon theoretical predictions.

In the above formula, the λ_c is maximized under the following conditions:

$$C_t \cong (1 + \gamma)^{-1}, \quad \gamma = (\lambda_{dt}/q_{1s}\lambda_{dt\mu})^{1/2}$$

In a D–T mixture, there are three molecules, D_2, DT, and T_2, with concentrations C_{D_2}, C_{DT}, and C_{T_2} determined by the conditions of chemical equilibrium. Thus, the rate $\lambda_{dt\mu}$ can be decomposed into the sum of two terms:

$$\lambda_{dt\mu} = \lambda_{dt\mu-d} C_{D_2} + \lambda_{dt\mu-t} C_{DT}$$

The idea of resonant molecular formation was experimentally confirmed qualitatively by the Dubna group in 1979 and in more detail by experiments at Los Alamos (Jones et al., 1983, 1986) and at PSI (Breunlich et al., 1987). In the latter experiments, both "three-body effects" exhibited in a density dependence of the density-normalized $\lambda_{dt\mu}$ and a strange temperature dependence were discovered. At the same time, a very rapid formation rate (of the order of $6 \times 10^8 \, s^{-1}$) was experimentally established for $\phi = \phi_0$ for a temperature range up to 500 K. These experimental data are summarized in Figure 5.9. Theoretical predictions based upon the resonant molecular formation model have not been able to explain the observed temperature dependence of the molecular formation rate; according to theoretical predictions, there should be a steeper decrease in $\lambda_{dt\mu}$ towards the lowest temperatures.

Experimentally, the existence of "triple collisions" in $\lambda_{dt\mu-d}$ has been consistently confirmed. The resonant reaction process between $(t\mu)$ and D_2 proceeds under the influence of a second D_2 molecule (Menshikov and Ponomarev, 1986), i.e.:

$$(t\mu) + D_2 + D_2 \rightarrow [(dt\mu)d2e]^* + D_2$$

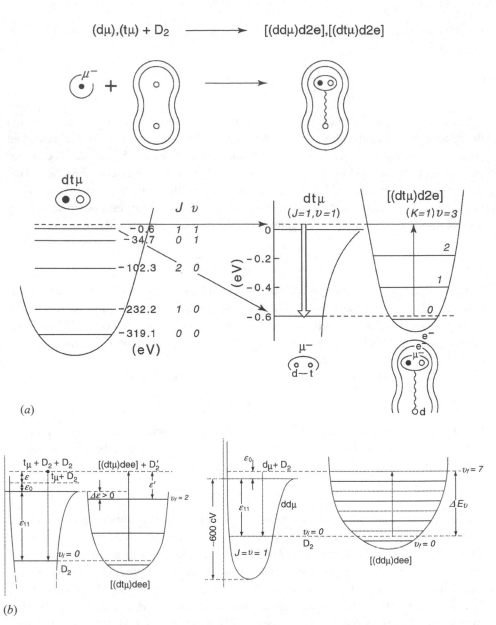

Figure 5.8 (*a*) Conceptual view of the resonant molecular formation mechanism of (dtμ) originally proposed by Gerstein and Ponomarev (1977). (*b*) Details of the revised energy-level diagram for resonant molecular formation (Ponomarev, 1990); dtμ, where one D_2 molecule, with a participation of the other D_2 molecule, is excited to $v_f = 2$ vs ddμ, where D_2 molecule is excited from ($v_i = 2$, $K_i = 2$) to ($v_f = 7$, $K_f = 1$).

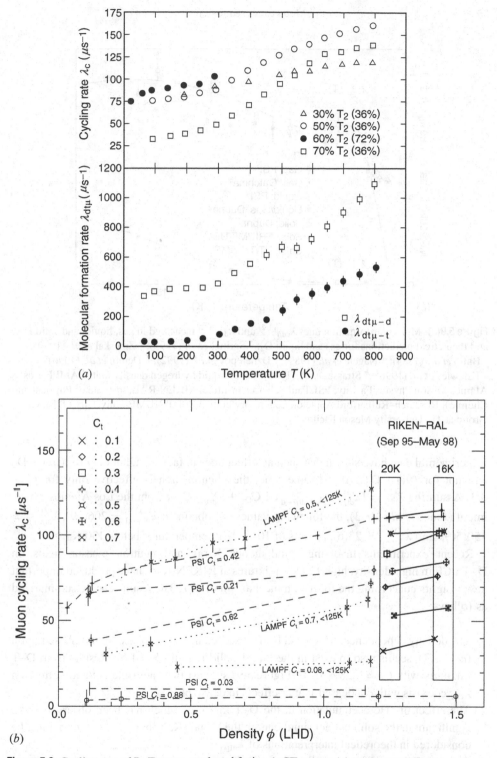

Figure 5.9 Cycling rate of D–T muon catalyzed fusion (µCF) versus (*a*) temperature (Jones *et al.*, 1983 and 1986) and (*b*) density (Kawamura *et al.*, 2003). LHD, liquid hydrogen density unit; LAMPF, Los Alamos Physics Facility; PSI, Paul Scherrer Institute; RIKEN-RAL, Institute of Physical and Chemical Research–Rutherford Appleton Laboratory branch.

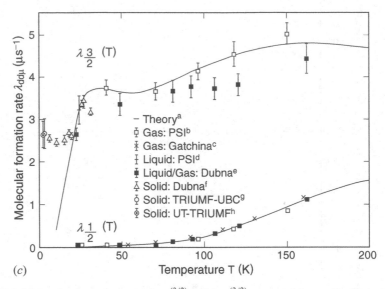

Figure 5.9(c) Molecular formation rates $\lambda_{dd\mu}^{(3/2)}$ and $\lambda_{dt\mu}^{(3/2)}$ measured in gas, liquid, and solid D_2, and theoretical predictions for gas and liquid D_2. [a] Scrinzi et al. (1993); [b] Zmeskal et al. (1990); [c] Balin et al. (1984); [d] Nägele et al. (1989); [e] Dzhelepov et al. (1992); [f] Demin et al. (1996); [g] Knowles et al. (1996); [h] Strasser et al. (1996). LHD, liquid hydrogen density unit; LAMPF, Los Alamos Meson Physics Facility; PSI, Paul Scherrer Institute; RIKEN-RAL, Institute of Physical and Chemical Research–Rutherford Appleton Laboratory branch; UT-TRIUMF, University of Tokyo Group at Tri-University Meson Facility.

Experimental data have shown that such an effect does in fact exist, but only for $(t\mu) + D_2$ and not for $(t\mu) + DT$. At the same time, the phenomenon is effective only for $C_t \geq 0.3$. Assuming $\lambda_{dt\mu} = [\lambda_{dt\mu-d}^{(1)} + \lambda_{dt\mu-d}^{(2)} \phi] C_{D_2} + \lambda_{dt\mu-t}^{(1)} C_{DT}$, from the Los Alamos experiment (Jones et al., 1983), the following values are obtained: $\lambda_{dt\mu-d}^{(1)} = 206(29)$, $\lambda_{dt\mu-d}^{(2)} = 450(50)$, and $\lambda_{dt\mu-d}^{(1)} = 23(6)$, in units of 10^6 s^{-1}, at temperatures below 130 K.

Recent experiments involving simultaneous X-ray and neutron measurements on D–T μCF in high-density, high-C_t D–T mixtures at RIKEN-RAL have produced important new insights concerning the formation mechanism of $dt\mu$. The results can be summarized as follows:

1. The density dependence, which had been observed in the gas phase–liquid phase region ($\phi = 1.2$), seems also to exist in the liquid–solid region ($\phi = 1.5$). Results from D–T mixtures with $C_t = 0.28 \sim 0.70$ (Figure 5.9) suggest that the triple collision effect on $\lambda_{dt\mu}$ coexists with condensed-matter effects in dense phases.
2. The effect of ^{3}He accumulation in the D–T mixture, which has been observed to be significant in the solid but not significant in the liquid (Kawamura et al., 1999), must be considered in theoretical interpretations of $\lambda_{dt\mu}$.
3. In systematic studies on μCF in solid D–T mixtures under a variety of conditions, i.e., tritium concentrations from 20% to 70%, and temperatures from 5 K to 16 K, an

increase in the muon cycling rate (λ_c) was observed with increasing temperature, as shown later. The observed changes in λ_c seem to be consistent with the temperature dependence in the dtμ formation process for λ_c (Kawamura *et al.*, 2003).

Compared to the formation of (dtμ), the formation of (ddμ) can be more quantitatively explained by theoretical predictions. Actually, the idea of resonant molecular formation was originally suggested by Vesman in relation to the enhanced (ddμ) formation rate (Vesman, 1967), and the resonant formation process is sometimes referred to as the Vesman mechanism. Thus, overall agreement between theory and experiment has been achieved for D_2 μCF but is not yet complete for D–T μCF. This general tendency can be qualitatively explained by considering the energy balance between the resonating muonic molecular state and the hydrogen molecule. As depicted in Figure 5.8, an energy deficiency exists for D_2 μCF, while an energy excess exists for D–T μCF so that one more collision partner is needed to take away the excess energy. As for the temperature dependence of $\lambda_{dt\mu}$, a phenomenological triple effect was introduced (Faifman, 1988), and pointed out that low-temperature data were able to be explained by shifting the transition energy from 0.601 eV to 0.596 eV.

A systematic experimental study on μCF was conducted using a series of solid deuterium and tritium mixtures. A variety of conditions were investigated, i.e., tritium concentrations from 20% to 70%, and temperatures from 5 K to 16 K. With decreasing temperature, a significant decrease in the muon cycling rate (λ_c) was observed (Figure 5.10b). The origins of the observed changes were interpreted by the temperature dependence in the dtμ formation process for λ_c (Kawamura *et al.*, 2003).

In the case of D_2 μCF, the ground state of (dμ) has two hyperfine states $F = \frac{3}{2}$ and $F = \frac{1}{2}$ and inelastic hyperfine transitions take place, such as $(d\mu)(F = \frac{3}{2}) + d \rightarrow (d\mu)(F = \frac{1}{2}) + d$. Because of spin dependence in (ddμ) molecular formation $\lambda_{dd\mu}{}^F$, the hyperfine transitions can be seen in a time-dependent change of fusion neutrons (Kammel *et al.*, 1983). As a result of the relatively slow intramolecular transition rate as well as slow fusion rate after the initial ddμ formation at (1, 1) state, the back-decay process ($[(dd\mu)d2e] \rightarrow d\mu + D_2$) competes with them. The temperature dependence in $\lambda_{dd\mu}{}^F$ is perfectly reproduced by the theory of resonant molecular formation below 200 K down to 25 K (liquid). As for the well-understood D_2 μCF, it has been found that there is a marked deviation between experiment and theory below 20 K, corresponding to μCF in the solid phase (Demin *et al.*, 1996; Knowles *et al.*, 1996), as summarized in Figure 5.9(c). The deviation is likely to originate in one or both of the following two mechanisms: (1) due to a nonthermalization effect during the slowing-down of (dμ), the existence of an energy gap in solid D_2 suppresses complete slowing-down, producing nonthermalized epithermal (dμ) (with energy 20 K); (2) the reaction mechanism of the resonant molecular formation process may be dramatically changed due to a change in the final-state energy spectrum, possibly including phonon excitation. Some theoretical work has been done (Adamczak and Faifman, 2001).

Understanding of muonic molecular formation can be somewhat extended by considering energetic (tμ) or (dμ). The cross-section for (dtμ) formation has been theoretically

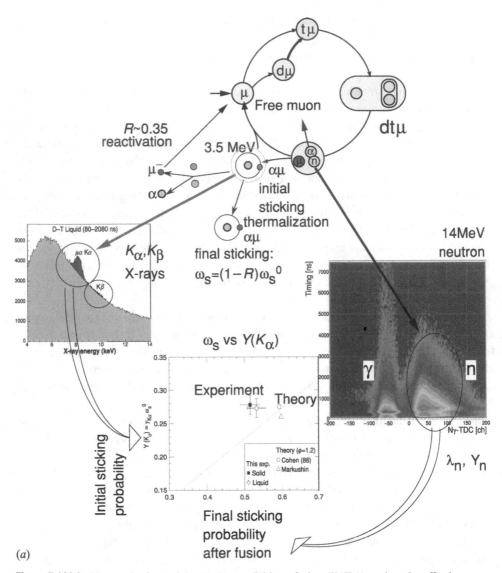

(a)

Figure 5.10(a) Observed values of the K_α X-ray yield per fusion ($Y(K_\alpha)$) against the effective sticking probability (ω_s) obtained from neutron data for the total loss rate. They are compared with the atomic process calculations on $\gamma(K_\alpha)(Y(K_\alpha) = \gamma(K_\alpha)\omega_s^0)$ and $R(\omega_s = (1 - R)\omega_s^0)$ with the assumption that initial sticking $\omega_s^0 = 0.912\%$. From Cohen (1988) *Phys. Rev.* **A38**, 44 and private communication (1998) and Markushin (1988).

calculated for various (tμ) energies and various energies (temperatures) of D$_2$ (Faifman and Ponomarev, 1991). At high energies, several significant resonances occur; for instance, (tμ) exhibits a strong resonance at 0.1 eV. An enhanced cross-section for (dtμ) formation can be expected compared to the cross-section for the elastic scattering which leads to the slowing-down, and so there is a possibility for resonances of this kind to be detected

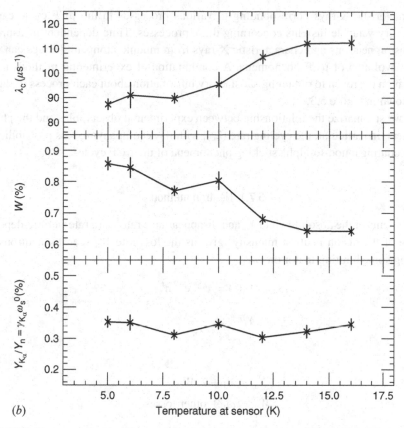

Figure 5.10(b) Temperature dependence of (top) the muon cycling rate (λ_c), (middle) the muon loss probability (W), and (bottom) the ratio of Y_{K_α} to Y_n in solid D–T with a tritium concentration of 40%.

experimentally. An experiment utilizing the Ramsauer effect in $H_2 + 0.1\%\,T_2$ to generate an energetic ($t\mu$) beam was carried out at TRIUMF (Marshall *et al.*, 1996).

As summarized in the review article by Ponomarev (1990), the formation rates in the systems other than ($dt\mu$) and ($dd\mu$) are mostly due to the nonresonant formation process of rotational–vibrational states (Jv). Theoretical values are on the whole consistent with experiments (presented for molecule (Jv)): theoretical rate (experimental rate) in $10^6\,\mathrm{s^{-1}}$); $pp\mu(10) - 2.2\,(2.5)$, $pd\mu(10) - 5.9(\sim 6)$, $pt\mu(10) - 6.5$(none), $tt\mu\,(11) - 3.0\,(2)$.

5.7 Muon sticking and regeneration in the μCF cycle

To date, a variety of experimental methods have been adopted in order to investigate μCF phenomena in D–T mixtures. Measurements of the 14-MeV fusion neutrons can be used to obtain the fusion neutron yield, the μCF cycling rate, and other parameters, usually with simultaneous decay e^- measurement for normalization purposes. Measurements of the characteristic X-rays corresponding to various processes in the μCF cycle can also provide very valuable insights concerning these processes. Time-dependent measurement of the fusion neutrons and characteristic X-rays from muonic atoms/molecules can reveal the time evolution of μCF phenomena. A combination of experimental methods may be the optimum approach to obtaining satisfactory information about each process of the μCF cycle shown in Figure 5.3.

Here we summarize the relationship between experimental observables and the physical parameters, in particular, the cycling rate (λ_c), with its accompanying loss probability (W) accommodating muon-to-alpha sticking phenomena of the μCF cycle.

5.7.1 Neutron method

Measurements of the absolute yield Y_n and disappearance rate λ_n (a rate of time-dependent decrease of the fusion neutron intensity) give us the loss rate W_n seen by neutrons, thus providing some limiting factor on ω_s.

$$Y_n(t) = \phi \lambda_c e^{-\lambda_n t}$$

$$Y_n = \frac{\phi \lambda_c}{\lambda_n}$$

where:

$$\lambda_n = \lambda_0 + \lambda_c W_n$$
$$W_n = \omega_s + \text{other losses}$$

5.7.2 X-ray method

X-ray measurements from $(\mu\alpha)^+$ ions give information directly related to sticking phenomena. Combining $Y_x(t)$ and $Y_n(t)$ leads to a direct measure of ω_s:

$$Y_x(t) = \phi \lambda_c \kappa \omega_s^0 e^{-\lambda_n t}$$

$$Y_x = \frac{\phi \lambda_c \kappa \omega_s^0}{\lambda_n} \quad \text{and} \quad Y(K_\alpha, K_\alpha \cdots) = Y_x/Y_n = \kappa(K_\alpha, K_\beta \cdots)\omega_s^0$$

where κ, given by the theory of the atomic processes of the $(\mu\alpha)^+$ ion, is the X-ray yield per sticking, and ω_s^0 is the initial sticking probability immediately after the fusion reaction in the muonic molecule. Actually, ω_s^0 is the sum of initial sticking contributions corresponding to each orbital of the $(\mu\alpha)^+$ ion:

$$\omega_s^0 = \sum_{n\ell} \omega_s^0(n\ell)$$

The ω_s which appears in the total loss probability W_n is obtained by correcting ω_s^0 with the regeneration factor R:

$$\omega_s = \omega_s^0 (1 - R)$$

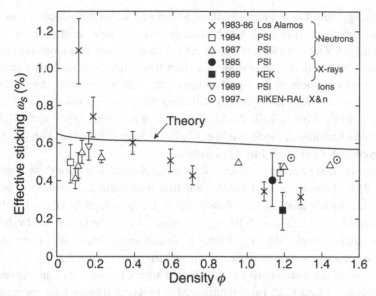

Figure 5.11 Existing data concerning the sticking probability ω_s vs density φ in muon catalyzed fusion in a D–T mixture.

Again, since the regeneration process depends upon the initial state of the $(\mu\alpha)^+$ ion, ω_s should be written as:

$$\omega_s = \sum_{n\ell} [1 - R(n\ell)]\, \omega_s^0 (n\ell)$$

The experimental values so far obtained through the loss rate W_n seen in the neutron method, together with the corrections due to other loss rates, are summarized in Figure 5.11, where values of the effective sticking, ω_s, are presented as a function of ϕ.

Theoretical studies of α-sticking were initiated by Jackson (1957), employing the sudden approximation. The probability of $(\mu\alpha)^+$ atom formation in an $n\ell$ state is given by:

$$\omega_s^0(n\ell) = \sum_m |F(n\ell m)|^2$$

where:

$$F(n\ell m) = \int \phi_{n\ell m}{}^*(r)\mathrm{e}^{-i\mathbf{q}\cdot\mathbf{r}}\psi_{\text{in}}(r)\mathrm{d}\mathbf{r}$$

where $\phi_{n\ell m}(r)$ is the wave function of $(\mu\alpha)_{n\ell m}$, and $q = m_\mu v$, where v is the velocity of $(\mu\alpha)$. In this expression, $\psi_{\text{in}}(r)$ is the normalized muon wave function at the instant of fusion, and can be expressed in terms of the muon molecule wave function, $\psi_{jv}(r, \mathbf{R})$, as $\psi_{\text{in}}(r) = N\psi_{jv}(r, \mathbf{R} = 0)$, where $\mathbf{R}$ is the internuclear distance and r is the muon coordinate with respect to the center of mass of the two nuclei, with N being the normalization constant. More refined theories correcting additional deficiencies in the existing theories have been proposed (Kamimura, 1989).

Since 1986, X-ray measurements have been applied several times to the direct measurement of the μ-α sticking probability in D–T μCF. As for X-ray detection in dtμ-μCF, the radiation background due to bremsstrahlung associated with t beta-decay is serious; this background, the energy range of which extends up to 17 keV, masks all the ω_s-related X-rays ($E(K_\alpha)$: 8.2 keV, $E(K_\beta)$: 9.6 keV, etc.). The use of pulsed muons is helpful here; by only operating the detection system in a short time interval around each muon pulse, a significant improvement in the signal-to-noise ratio can be obtained. An experiment at PSI (Bossy et al., 1987) was performed with continuous muons upon a low C_t ($\sim 10^{-4}$) D–T mixture, while those at KEK-MSL (Nagamine et al., 1987, 1990) and RIKEN-RAL (Nagamine and Kamimura, 1998; Ishida et al., 2003) were carried out with pulsed muons and employed high C_t (from 0.1 to 0.7) mixtures.

Following the first successful observation of K_α X-rays from $(\mu\alpha)^+$ in high-ϕ high-C_t ($C_t = 0.3$) D–T mixtures at KEK-MSL, the first systematic data on both ω_s and $Y(K_\alpha)$ were subsequently obtained in high-density ($\phi = 1.2$–1.5) high-C_t ($C_t = 0.1$–0.7) D–T mixtures at RIKEN-RAL (Figure 5.10(a)). The experimental results provided values for (1) the effective sticking probability ω_s obtained by fusion neutron data (after correction for all the other loss processes such as those in ddμ and ttμ) and (2) the K_α X-ray yield $Y(K_\alpha)$ (photon/fusion) of the recoiling $(\alpha\mu)^+$ ion formed after a μ^--to-α sticking process.

The results for ω_s and $Y(K_\alpha)$ are summarized in Figure 5.10(a), where our results on ω_s and $Y(K_\alpha)$ are compared with the atomic process calculation (Markushin, 1988; Struensee et al., 1988). Here we used a theoretical value of $\omega_s^0 = 0.912\%$ (Hu et al., 1994) for the initial sticking probability. It can be seen that our ω_s is smaller than any theoretical values so far published, while the discrepancy in the $\alpha\mu$ X-ray yield seems to be less significant. In addition, the measured $Y(K_\beta)/Y(K_\alpha)$ values (0.075 ± 0.010 for liquid, 0.060 ± 0.012 for solid; Nakamura et al., 2000) are much smaller than the calculated values ($\cong 0.12$) with correct Stark mixing at $n = 3$.

The result suggests that the initial sticking probability should be smaller ($\omega_s^0 \sim 0.75\%$) if we believe the atomic process calculation. Another possible explanation is, if we believe the ω_s^0 calculation, that the atomic process of $(\alpha\mu)^+$ ions has to be reconsidered to explain ω_s, $Y(K_\alpha)$ and $Y(K_\beta)/Y(K_\alpha)$ consistently. Considering the smallness of the observed K_β/K_α intensity ratio, it is likely that the excitation rates from the $n \geq 3$ levels are higher than calculated. This would lead to smaller ω_s, while the $Y(K_\alpha)$ would not be much affected. Thus, although there have been no quantitative explanations, the existence of "anomalous" regeneration (ionization of the recooling $(\alpha\mu)^+$ ion) has been experimentally observed in high-density high-C_t D–T mixtures.

Furthermore, experimental investigation was made at RIKEN-RAL on the D–T μCF in solid D–T with $C_T = 0.40$ in the temperature range from 5 K to 20 K (Kawamura et al., 2003). A significant change in the ω_s seen via a loss of the fusion neutron against temperatures was found (Figure 5.10b). Based upon observations of no change of Y_x/Y_n, the origin of the observed changes was interpreted as due to temperature dependence in the muon reactivation process after muon-to-alpha sticking influencing W; there were anomalously small ω_s (0.42) and large R at 16 K, and larger ω_s (0.61) and smaller R at 5 K, both of which are rather consistent with theories. This anomalous regeneration is an interesting objective for further theoretical investigation. At the same time, it may encourage the development of an idea for the enhancement of energy production in D–T μCF. The result suggests that a further smaller ω_s with larger R can be obtained in a solid at high temperature.

5.8 Application to energy sources and neutron sources

Possible applications of the μCF phenomena have been considered in various fields. Currently, a realization of the following three subjects is frequently discussed: (1) an energy source; (2) a 14-MeV neutron source; and (3) an ultraslow μ^- source. Since the last subject was described in Chapter 2, let us consider here the first two subjects.

5.8.1 A practical energy source using μCF

Before entering into the consideration of energy production efficiency, in particular, regarding the possibility of break-even achievement, we must emphasize the important features of the μCF in the energy problem. As commonly mentioned for nuclear fusion energy, the μCF is a "clean" energy source with a production of "minimal" radioactive waste. In contrast to thermonuclear fusion, because of the use of the heavy electron of the μ^-, there is no need to use very high temperatures; the μCF is a real "cold" fusion. μCF can only take place with the introduction of the accelerator producing μ^-, it can be stopped at the instance of the termination of the accelerator operation; the μCF energy production is therefore a "controlled" atomic energy source.

In order to consider energy production efficiency, it is required to know how much energy is needed to produce a single muon (the muon cost). There have been several discussions on the optimization of the π^- production and $\pi^- \rightarrow \mu^-$ conversion processes. In the case of π^- production, the fundamental reactions in the nucleon–nucleon inelastic process are $n + n \rightarrow p + n + \pi^-$ and $n + p \rightarrow p + p + \pi^-$. Therefore, for efficient μ^- production, the use of accelerated nuclei which contain "accelerated" neutrons is inevitable. Energy (cost) estimates towards economical μ^- production have been carried for deuteron and triton beams.

Following the argument made by Petrov et al. (1979), a realistic solution seems to be as follows. Using a 1-GeV/nucleon t(d) beam to bombard Li or Be nuclei, we can obtain 0.22(0.17) π^- from a single t(d). With the use of a large-scale superconducting solenoid with a reflecting mirror, one can expect 75% efficiency for μ^- production from a single π^-. Since π^- production is proportional to the incident t(d) beam energy, and 1 GeV produces 0.17 μ^-, one μ^- per d(t) can be produced using an energy of 6(8) GeV. Selecting the corresponding values for π^- production in a t-t collision, the eventual cheapest cost might be about 1 π^-/4 GeV and 1 μ^-/5 GeV.

Several ideas have been proposed for the reduction of the muon cost. Studies have been carried out to optimize the type of particle, the particle energy, and the choice of fixed target. To summarize, optimization does not seem to yield an improvement in the value mentioned above ($1\pi^-$/4 GeV). One possible method to reduce the π^- production cost is to use a colliding beam (Chapline and Moir, 1986). In this case, the energy of the center-of-mass motion, which is wasted in the case of a fixed-target geometry, would be used efficiently. It is claimed that a production efficiency of $1\pi^-$/1.8 GeV($0.55\pi^-$/GeV) can be achieved

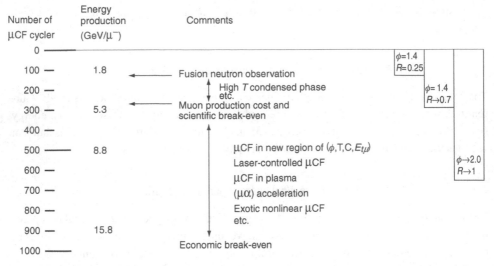

Figure 5.12 Number of fusions and energy produced from the D–T muon catalyzed fusion (μCF), and expected future increase under the projected changes in density (ϕ) and regeneration factor (R).

using a d-d collider. However, the feasibility of such a collider with a power exceeding MW is totally uncertain.

On the other hand, the energy production capability $E_{\mu CF}^{out}$ of the μCF process is determined by $E_{\mu CF}^{out} = 17.6 \times Y_n$ (MeV) in the case of D–T μCF, which has a stringent limiting factor due to the sticking probability ω_s; this can be expressed as $E_{\mu CF}^{out} \leq 17.6 \times \omega_s^{-1}$(MeV). The situation in relation to $E_{\mu CF}^{out}$ is summarized in Figure 5.12. The scientific break-even can be obtained for $E_{\mu CF}^{out} > 5$ GeV. As described earlier, based upon recent results at RIKEN-RAL, the most promising way to achieve the break-even seems to be the D–T μCF in a high T condensed phase such as at high-temperature solid, where enhanced $R(\rightarrow 0.7)$ might be obtained.

Several remarks can be made on the possibilities for a further increase in the energy production capability of D–T μCF towards economic break-even:

1. Since the conditions so far used for the D–T target in the μCF experiment, such as density, temperature and C_t, as well as the energy of the (tμ) atoms $E_{t\mu}$, have not been satisfactory, there may exist more favorable conditions for higher energy production; one possibility would be a μCF experiment with a higher-density D–T mixture, on the order of $\phi \cong 2\phi_0$.
2. In order to increase $\lambda_{dt\mu}$, a more favorable matching condition in terms of resonant molecular formation may exist which could be accessed by exciting the molecular levels of D_2 or DT using, e.g., lasers.
3. In order to decrease ω_s, or in order to increase R, several ideas have been proposed: among them, the use of a D–T plasma, where enhanced regeneration is expected due to an elongated $(\alpha\mu)^+$ mean-free path (Menshikov et al., 1989), and acceleration of $(\mu\alpha)^+$using an electric field (Daniel, 1990, 1991) are two that may be worth trying.

4. Due to collisions between the $(\alpha\mu)^+$ ions and α^{++} ions from nearby μCF reactions, exotic regeneration reactions may occur in high-density μCF in D–T mixtures with an intense pulsed μ^- beam; $(\mu\alpha)^+(I) + \alpha^{++}(II) \rightarrow \mu^- + \alpha^{++}(I) + (\alpha)^{++}(II)$.

For the purpose of exotic regeneration it is indispensable to realize a very intense muon channel like the "super-super muon channel" described in Chapter 2. With such a system, $2 \times 10^{11}\mu^-$/s can be obtained with 25 MeV/c $\leq$ muon momentum $\leq$ 120 MeV/c, and, with a pulsed accelerator, one can expect an instantaneously ultraintense muon beam of $4 \times 10^9\mu^-$/pulse. Using intense pulsed μ^- of this kind, μCF phenomena with ultrahigh fusion density can hypothetically be realized on a scale such as 3×10^{13} fusions/s in 5ℓ target volume. The instantaneous μ^- intensity per unit volume of $2 \times 10^8\mu^-$/pulse per cc, which corresponds to a μCF density of $3 \times 10^{10}\mu$CF/pulse per cc, might be sufficient to yield interesting nonlinear μCF phenomena.

In the experiment at RIKEN-RAL, the level of energy production is 3 μW((6-7) $\times$ $10^3\mu^-$/s produced by the superconducting muon channel installed at 800 MeV $\times$ 200 μA proton beam with 10 mm carbon target and stopping in 1.5 cc liquid/solid D–T target, producing 10^6 fusion/s). The basic understanding of the μCF phenomena and conceptual ideas of the μCF energy source will make further progress as a result of the progress of the high-intensity hadron accelerator as well as the advanced muon-producing beam chan-nel. The new accelerator projects, like Spallation Neutron Source, Neutrino Factory, Muon Colliders, may contribute significantly to such progress. Even with the presently available value of the fusion yield of close to 150 per one μ^-, by employing advanced methods of either accelerator or muon production, one can realize the μCF reactor at the levels of kW to MW. Such situations are summarized in Figure 5.13. For the overall development of fusion energy, it is quite important to make a public demonstration of the realization of kW fusion energy production by using the μCF in the near future.

Considering these new trends related to μCF processes, one might expect a contribution to the development of fusion energy. Some prominent examples can be summarized as follows: (1) materials development for the innermost wall of the proposed fusion reactor, using a high flux of 14 MeV neutrons from the μCF, as described later; (2) tritium production test facility, where, by using a nature of high spatial fusion density, the tritium breeding process at the blanket proposed for the plant of fusion reactor can be examined more easily; (3) contribution to the studies of plasma instability due to alpha-heating, where application of the μCF process adjacent to a plasma facility can be used in the study of selected aspects of the instability phenomena.

As opposed to the production of energy solely via μCF, the concept of a muon catalyzed hybrid reactor has been proposed by Petrov (1980) and later by Eliezer et al. (1987). In this concept, the accelerated 1 GeV/nucleon d beam is used to bombard a Li or Be target, with the remainder of the beam stopping in ^{238}U; in this way $\sim$30% of the beam is spent on π^- production, while 70% is spent on ^{238}U fission and ^{238}Pu production via electronuclear breeding. The π^- thus produced is used for μCF in a D–T mixture, and the 14 MeV neutron produced in the μCF reaction stops in the blanket of ^{238}U and ^{6}Li, yielding ^{239}Pu and T. The Pu fuels a thermal nuclear reactor, and the fission energy is used to feed the accelerator

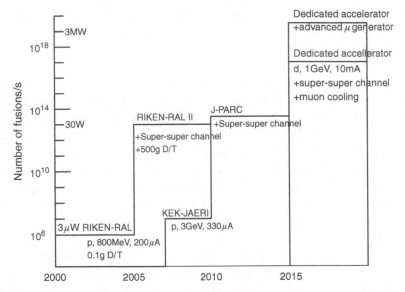

Figure 5.13 Expected increase in the power generation of the D–T muon catalyzed fusion (μCF) against expected progress in accelerator power and methodology of muon beam production. Here, a fusion rate of $150/\mu^-$ is assumed. RIKEN-RAL, Institute of Physical and Chemical Research–Rutherford Appleton Laboratory branch; KEK-JAERI, High Energy Accelerator Research Organization–Japan Atomic Energy Research Institute, J-PARC, Japan Proton Accelerator Research Complex.

and the rest of the system. The proposers conclude that the hybrid system can double the electric power output of the nonhybrid electronuclear breed. However, there is an argument against the use of μCF for fuel production in a thermal nuclear reactor, which brings all the usual problems of nuclear reactors such as radioactive waste disposal.

5.8.2 14 MeV neutron source using μCF

When thermal nuclear fusion becomes realistic, it is important to develop materials to be used for the first wall immediately adjacent to the innermost core of the thermal fusion reactor. For this purpose, it is necessary to have an irradiation test facility where candidate materials can be tested under a very high flux of 14 MeV neutrons. One practical idea is to have an intense 200 keV d beam as primary beam and produce 14 MeV neutrons via the $d + t \rightarrow \alpha + n$ reaction. In parallel to this idea, the 14 MeV neutrons from μCF can be considered to be an alternative plan for such a materials irradiation facility.

Some promising schemes have been considered (Petitjean *et al.*, 1993; Anissimov *et al.*, 2001). Let us assume that a 2 GeV 12 mA deuteron accelerator becomes available. By placing a 150-cm-long and 0.75-cm-radius lithium target under the confinement field of a 17/7 T superconducting solenoid, intense pion production and efficient μ^- production can be realized. In this scenario, μCF occurring in a D–T target into which the intense μ^- beam is introduced produces intense 14 MeV neutrons in quantities of the order of 10^{14} n/cm^2 per s in the test volume of $\sim 2.5\ell$ with a surface of ~ 350 cm^2, and the material to be tested is placed on one surface of the D–T container. Most importantly, power consumption by the μCF method is substantially lower compared to that in the d accelerator method. Some realistic plant design work is now in progress.

5.9 Present understanding and future perspectives

At the beginning of the twenty-first century, our present understanding of μCF together with future perspectives on fusion energy will be summarized below. Let us again confine our considerations to D–T μCF.

1. Some important experimental facts on D–T μCF have not yet been explained at all, in particular, for μCF in the high-density and condensed-matter phase of the D–T mixture. Distinguished examples are the rate of muon molecule formation $\lambda_{dt\mu}$ in liquid as well as solid D–T mixture (while the theory predicts a reduced value of $\lambda_{dt\mu}$, the experiment has shown the highest one) and the anomalously large regeneration factor that exists in the μCF in liquid and solid D–T mixture for the $(\alpha\mu)^+$ formed in the fusion process in the $(dt\mu)$ molecule.

2. A realization of the scientific break-even of 300 fusions per single μ^- seems to be taking place in the near future. The presently realized numbers of (fusion numbers/μ^-, effective sticking probability ω_s, regeneration factor R) are $(150/\mu^-, 0.44\%, 0.52)$ and can be replaced by numbers such as $(300/\mu^-, 0.33\%, 0.70)$, at least in the following two schemes: (1) by extrapolating the anomalous temperature dependence in R, one can expect enhanced R in pressurized and high-temperature (above 20 K) solid D–T; (2) by introducing high-intensity pulsed μ^-, one can expect a nonlinear effect of correlated fusion phenomena to enhance R.

3. Large-scale μCF set-up, if realized, may contribute to the development of fusion energy research and development. By employing high-intensity muon generation, such as described in Chapter 2, one can realize fusion phenomena with extremely high spatial density. Thus, the μCF plant can be used as (1) intense 14 MeV neutron source for the development of wall materials for the fusion reactor and (2) tritium fuel production.

4. Realization of the μCF reactor at the level of kW at the intense hadron accelerator now or in the near future to demonstrate the practicability of continuous production of fusion energy is urgently required to enhance public understanding of fusion energy.

REFERENCES

Ackerbauer, P. *et al.* (1993). *Hyperfine Interactions*, **82**, 357.
Ackerbauer *et al.* (1999). *Nucl. Phys.*, **A652**, 311.
Adamczak, A. and Faifman, M.P. (2001). *Phys. Rev.*, **A64**, 052705.
Adamczak, A. and Melezhik, V.S. (1988). *Muon Catalyzed Fusion*, **2**, 131.
Alvarez, L.W. *et al.* (1957). *Phys. Rev.*, **105**, 1127.
Anissimov, V.V. *et al.* (2001). *Fusion Technol.*, **39**, 198.
Balin, V.D. *et al.* (1984). *Phys. Lett.*, **B41**, 173.
Baumann, P. *et al.* (1987). *Muon Catalyzed Fusion*, **1**, 87.
Bogdanova, L.N. *et al.* (1982). *Sov. Physics, JETP*, **56**, 931.

Bogdanova, L.N. *et al.* (1988). *Muon Catalyzed Fusion*, **3**, 359.

Bogdanova, L.N. *et al.* (1990/91). *Muon Catalyzed Fusion*, **5/6**, 189.

Bossy, H. *et al.* (1987).*Phys. Rev. Lett.*, **59**, 2864.

Breunlich, W.H. *et al.* (1987). *Muon Catalyzed Fusion*, **1**, 121.

Breunlich, W.H. *et al.* (1987). *Phys. Rev. Lett.*, **58**, 329.

Breunlich, W.H. *et al.* (1989). *Annu. Rev. Nucl. Sci.*, **39**, *311*.

Chapline, G. and Moir, R. (1986). Lawrence Livermore National Laboratory Report.

Crawford, J. E. *et al.* (1991). *Phys. Rev.*, **D43**, 46.

Czaplinski, W. *et al.* (1994). *Phys. Rev.*, **A50**, 518 and 525.

Daniel, H. (1990/91). *Muon Catalyzed Fusion*, **5/6**, 335.

Demin, D.L. *et al.* (1996). *Hyperfine Interactions*, **101/102**, 13.

Dzhelepov, V.P. *et al.* (1966). *Sov. Phys. JETP*, **19**, 820.

Dzhelepov, V.P. *et al.* (1992). *Sov. Phys. JETP*, **74**, 589.

Eliezer, S. *et al.* (1987). *Nuclear Phys.*, **127**, 527.

Faifman, M.P. *et al.* (1988). *Muon Catalyzed Fusion*, **2**, 285.

Faifman, M.P. and Ponomarev, L.I. (1991). *Phys. Lett.*, **B265**, 201.

Frank, F.C. (1947). *Nature*, **160**, 525.

Friar, J.L. *et al.* (1991). *Phys. Rev. Lett.*, **66**, 1827.

Froelich, P. *et al.* (1995). *Phys. Rev. Lett.*, **75**, 2108.

Gerstein, S.S. and Ponomarev, L.I. (1977). *Phys. Lett.*, **728**, 80.

Hale, G. M. (1990/91). *Muon Catalyzed Fusion*, **5/6**, 227.

Hu, C.-Y. *et al.* (1994). *Phys. Rev.*, **A49**, 4481.

Ishida, K. *et al.* (2003). *Phys. Rev.*, submitted.

Jackson, J.D. (1957). *Phys. Rev.*, **106**, 330.

Jones, S.E. *et al.* (1983). *Phys. Rev. Lett.*, **52**, 1757.

Jones, S.E. *et al.* (1983). *Phys. Rev. Lett.*, **56**, 588.

Kamimura, M. (1989). *AIP Conference Proc.*, **181**, 330.

Kammel, P. *et al.* (1983). *Phys. Rev.*, **28A**, 2611.

Kawamura, N. *et al.* (1999). *Phys. Lett.*, **B465**, 74.

Kawamura, N. *et al.* (2003). *Phys. Rev. Lett.*, **90**, 043401-1.

Knowles, P.E. *et al.* (1996). *Hyperfine Interactions*, **101/102**, 21.

Kravtsov, A.V. *et al.* (1984). *JETP Lett.*, **40**, 875.

Lauss. B. *et al.* (1996). *Phys. Rev. Lett.*, **76**, 4963.

Markushin, V.E. (1988). *Muon Catalyzed Fusion*, **3**, 395.

Matsuzaki, T. *et al.* (1999). *Hyperfine Interactions*, **119**, 361.

Matsuzaki, T. et al. (2002). *Nucl. Instruments*, **A408**, 814.

Marshall, G.M. *et al.* (1990). *Proc. Int. Sympo. on Muon Catalyzed Fusion* (RAL-90-022).

Marshall, G.M. *et al.* (1996). *Hyperfine Interactions*, **101/102**, 47.

Menshikov, L.I. and Ponomarev, L.I. (1986). *Phys. Lett.*, **167B**, 141.

Menshikov, L.I. *et al.* (1989). *Sov. Phys. JETP*, **68**, 258.

Nagamine, K. and Kamimura, M. (1998). *Adv. Nucl. Phys.*, **24**, 151.

Nagamine, K. *et al.* (1987). *Muon Catalyzed Fusion*, **1**, 137.

Nagamine, K. *et al.* (1990). *Muon Catalyzed Fusion*, **5**, 239.

Nägele, N. *et al.* (1989). *Nucl. Phys.*, **A493**, 397.

Nakamura, S.N. *et al.* (2000). *Phys. Lett.*, **B473**, 226.

Petitjean, C. *et al.* (1990/91). *Muon Catalyzed Fusion*, **5/6**, 199.

Petitjean, C. *et al.* (1993). *PSI Report*, **PSI-PR-93-09**.

Petrov, Y.V. *et al.* (1979). *Sov. J. Nucl. Phy.*, **30**, 66.

Petrov, Y.V. (1980). *Nature*, **285**, 466.

Phillips, A.C. *et al.* (1975). *Nucl. Phys.*, **A237**, 493.

Ponomarev, L.I. (1983). *Atomkernenerg./Kerntechnik.*, **43**, 3.

Ponomarev, L.I. (1990). *Contemporary Phys.*, **31**, 219.

Ponomarev, L.I. *et al.* (1996). *JETP Lett.*, **64**, 139.

Sakamoto, S. *et al.* (1996). *Hyperfine Interactions*, **101/102**, 297.

Sakharov, A.D. (1948). *Report FIAN*, **1**.

Scrinzi, A. *et al.* (1993). *Phys. Rev.*, **A47**, 4691.

Strasser, P. *et al.* (1996). *Phys. Lett.*, **B368**, 32.

Struensee, M.C. *et al.* (1988). *Phys. Rev.*, **A37**, 340.

Szalewicz, K. *et al.* (1990/1991). *Muon Catalyzed Fusion*, **5/6**, 241.

Vesman, E.A. (1967). *JETP Lett.*, **5**, 91.

Vinitsky, S.I. *et al.* (1979). *Soviet Phys. JETP*, **47**, 444.

Vinitsky, S.I. *et al.* (1982). *Soviet Phys. JETP*, **55**, 578.

Voropaev, N. J. *et al.* (2001). *Hyperfine Interactions*, **138**, 331.

Yukhimchuk, A. *et al.* (1999). *Hyperfine Interactions*, **119**, 361.

Zaplinski, W. *et al.* (1994). *Phys. Rev.*, **A50**, 518, 525.

Zavattini, E. (1975). In: *Muon Physics*, vol. 2, ed. V.W. Hughes and C.S. Wu, p. 219. New York Academic Press.

Zmeskal, J. *et al.* (1990). *Phys. Rev.*, **A42**, 1165.

6

Muon spin rotation/relaxation/resonance:
basic principles

The principle of the muon spin rotation/relaxation/resonance (μSR) method is based upon the laws of particle physics. As seen in Figure 1.5, the spin of the $\mu^+(\mu^-)$, when it is born via the decay of the $\pi^+(\pi^-)$, is completely polarized along the direction of its motion; once the $\mu^+(\mu^-)$ are focused or collimated along one direction, the resulting beam is polarized along its direction of motion. During the slowing-down of the $\mu^+(\mu^-)$ inside the host material, as described in Chapter 3, the spin polarization is maintained in the long-lived form of diamagnetic μ^+, paramagnetic Mu (ortho state with spin $= 1$), or ground state of a muonic atom in the case of μ^-. After stopping at some specific microscopic location, the $\mu^+(\mu^-)$ decays into $e^+(e^-)$ and two neutrinos, as shown in Figure 1.5, with the $e^+(e^-)$ spatial distribution oriented preferentially along the $\mu^+(\mu^-)$ spin direction. The decay $e^+(e^-)$ energy ranges up to 50 MeV, and the direction of the $\mu^+(\mu^-)$ spin can be observed in a time-resolved fashion by measuring these high-energy $e^+(e^-)$ using detectors placed outside the target material to be investigated; measurements are carried out under variations of external conditions such as temperature, pressure, and applied magnetic or electric fields.

The μSR method can be considered as a sensitive magnetic "compass" to probe the microscopic magnetic properties of condensed matter. As will be recognized in detail in Chapters 7 and 8, the points of advantage of the μSR method, in comparison with other microscopic probes such as neutron scattering, synchrotron radiation, or nuclear magnetic resonance (NMR), can be summarized as follows:

1. Because the spin polarization is provided by the laws of particle physics, microscopic magnetic properties can be studied under zero external field; this is a significant advantage for studies of, for example, superconductors.
2. Sensitive microscopic field measurements with highly efficient detection of the radioactive decay product can be realized by employing high flux of the μ^+ beam where 10^6 decay e^+ events/min can be easily obtained from a sample of less than 100 mg.
3. With the help of the muon's large magnetic moment (3.2 times that of the proton) and relevant lifetime, the μSR method is sensitive to very weak (down to less than a Gauss) and randomly oriented microscopic magnetic fields.
4. Again, as a result of the value of the magnetic moment and of the time window imposed by the muon lifetime, μSR is sensitive to the dynamics of surrounding electronic spins

(a) Muon spin rotation (b) Muon spin relaxation (c) Muon spin resonance

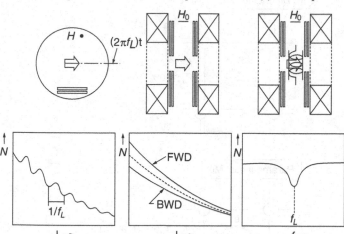

Figure 6.1 Schematic view of muon spin rotation/relaxation/resonance (μSR) experimental arrangements for (a) spin rotation, (b) spin relaxation, and (c) spin resonance experiments. External field, muon spin direction, and counter geometry are indicated.

which have a characteristic correlation time somewhat slower than the sensitivity range of neutron scattering and somewhat faster than that of NMR.

There are two primary ways of classifying μSR methods: (1) according to experimental arrangement; and (2) according to the type of muon state. As summarized in Figure 6.1, depending upon the geometrical configuration of external field and/or decay e^+/e^- counters (with respect to the initial muon spin direction), there are three types of μSR method: (1) muon spin rotation; (2) muon spin relaxation; and (3) muon spin resonance. Also, as summarized in Figure 6.2, depending upon the state of the spin-polarized muon to be used, we have three types of μSR method: μ^+SR, MuSR, and μ^-SR.

Historically, the development of μSR was initiated by the experiment on discovery of parity violation in the decay of polarized muons (Friedman and Telegdi, 1957; Garwin et al., 1957), which was followed by various pioneering experimental and theoretical studies of weak interaction processes. The principles and condensed-matter applications of μSR are reviewed in various monographs (distinguished examples: Schenck, 1985; Karlsson, 1995; Schatz and Weidinger, 1996; Sonier, private communication, 2002) and proceedings of regular international conferences (most recently, Nagamine et al., 1997; Roduner et al., 2000; Heffner et al., 2003).

6.1 Muon spin rotation

Now let us apply an external field H_0 perpendicular to the initial muon spin direction. The muon, after stopping inside the stopping material, takes spin precession around H_0 and the time spectrum of electron / positron observed at angle (θ) with respect to the beam direction is

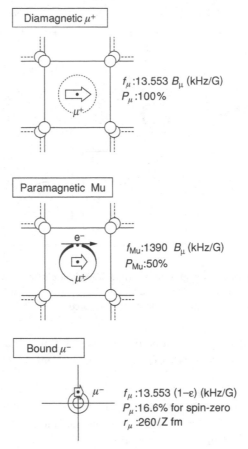

Figure 6.2 Schematic picture of diamagnetic μ^+, paramagnetic Mu, and bound μ^- of the ground state of muonic atoms in condensed matter with summaries of the basic properties of these states relevant to muon spin rotation/relaxation/resonance (μSR) studies.

expressed by:

$$N(\theta, t) = N_0 \exp(-t/\tau)[1 + A \cos (\theta - \omega_0 t)], \; \omega_0 = \gamma_\mu H_0$$

where the gyromagnetic ratio of the muon, γ_μ, is given in terms of the muon magnetic moment:

$$\gamma_\mu = g_\mu e h / 2 m_\mu c$$

The sign and magnitude of A, as described in Chapter 1, depend on: (1) μ^+ or μ^-; (2) the method of beam production (nature of beam channel); (3) beam polarization; (4) how to average energy spectrum in e^+/e^- counters. Roughly speaking, for backward-decay μ^+ beam $A \cong +\frac{1}{3}$, for surface μ^+ beam $A \cong -\frac{1}{3}$, and for backward-decay μ^- beam and μ^- bound to a zero-spin nucleus with a change of lifetime $A \cong +\frac{1}{18}$.

The following convenient numerical relationships can be obtained between the precession frequency of the muon spin and the local field H_μ at the muon site:

$$F_\mu(\text{kHz}) = 13.553 \times H_\mu(\text{G}) \qquad \text{for } \mu^+$$

$$f_{\text{Mu}}(\text{kHz}) = 1390 \times H_\mu(\text{G}) \qquad \text{for free muonium}$$

$$f_\mu(\text{kHz}) = 13.553(1 - \varepsilon) \times H_\mu(\text{G}) \quad \text{for bound } \mu^-$$

In the above equations, in nonmagnetic materials the H_μ for μ^+ is closely related to H_0, with some material-dependent corrections like Knight shift, paramagnetic shift, while, in magnetically ordered material, H_μ is a characteristic hyperfine or internal field due to ordered magnetic spin, as described later. The H_μ for Mu in various materials is subject to substantial change from that for free Mu, due to the electronic state of Mu, as described later. The H_μ for bound μ^- at the ground state of muonic atoms around nuclei with atomic number Z is very similar to the nuclear hyperfine field of $(Z - 1)$ nuclei with some corrections, as described later, and in this case ε is correction of the μ^- magnetic moment (see section 4.1).

The experimental arrangement is shown schematically in Figure 6.1(a). In the normal case the muon is introduced with its polarization direction longitudinal to the beam direction, and the field is applied in the plane perpendicular to the muon beam axis. However, when a spin rotator (see section 2.1.5) is employed in order to produce a transversely polarized muon beam, the field is applied parallel to the beam direction.

In the presence of inhomogeneous field broadening in H_μ, the precession amplitude is not constant, but experiences damping with time, as expressed by:

$$N(\theta, t) = N_0 \exp(-t/\tau)[1 + AG_x(t) \cos(\theta - \omega_0 t)]$$

The damping function here, $G_x(t)$, is called the transverse relaxation function. In this static broadening case, it is the Fourier transform of the local field distribution $p(\Delta H)$:

$$G_x(t) = \int p(\Delta H) \cos[(\gamma_\mu \Delta H)t] d(\Delta H)$$

The transverse relaxation function also reflects dynamic fluctuations of the local field as a function of time, and so $G_x(t)$ alone cannot discriminate between static and dynamic origins.

A conceptual understanding of how random dynamical fields affect the muon spin relaxation rate λ_R can be obtained as shown in Figure 6.3. Suppose there is a time-dependent fluctuation of the muon local field which changes from $+|H_z|$ to $-|H_z|$ or vice versa within an interval τ_c. Then the precessional angular displacement of the muon spin ($\delta\phi$) changes from $+\gamma_\mu|H_z|\tau_c$ to $-\gamma_\mu|H_z|\tau_c$. Within a time t, the frequency of occurrence of these fluctuations is t/τ_c. Then, according to random walk theory, the average mean-squared deviation during the time period t is $\langle\delta\phi^2\rangle = (t/\tau_c)\delta\phi^2 = (t/\tau_c)(\gamma_\mu^2 H_z^2 \tau_c^2)$. The relaxation time T_μ, the time required for $\sqrt{\langle \delta\phi^2 \rangle}$ to equal 1 rad, is then as follows: $1 = (T_\mu/\tau_c)(\gamma_\mu^2 H_\zeta^2 \tau_c^2)$, and finally the following relation can be obtained:

$$\lambda_R = (T_\mu)^{-1} = \gamma_\mu^2 H_z^2 \tau_c$$

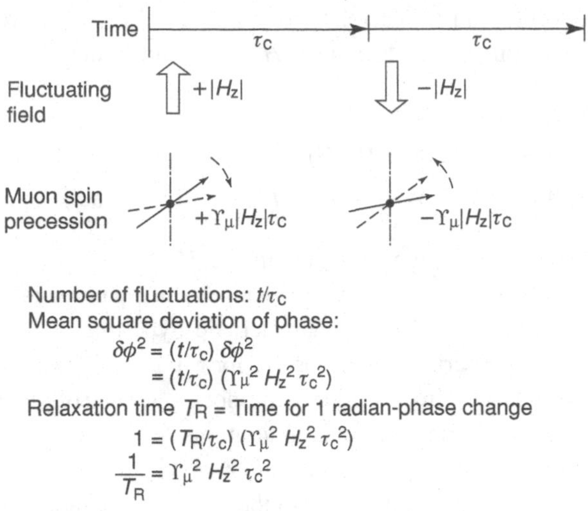

Figure 6.3 Conceptual view of the principle of spin relaxation. Due to the time-dependent orientationally random fluctuating field (in this case, up and down) sensed by the μ^+, the memory of the initial spin polarization is lost.

6.2 Muon spin relaxation

The initial muon spin may be relaxed when the muon senses the local field distribution and its dynamic fluctuation. When the local field has no preferential direction, the observable quantity is a relaxation of the longitudinal polarization; that is to say, the projection of the muon spin along its initial value $\sigma_z(t)$ experiences a time-dependent change:

$$G_z(t) = <\sigma_z(t)\sigma_z(0)>$$
$$N(\theta, t) = N_0 \exp(-t/\tau_\mu)[1 + AG_z(t)\cos\theta]$$

This function can be experimentally observed from forward/backward asymmetry, as shown in Figure 6.4 in the case of a pulsed muon beam; this is measured by setting a counter in each of the forward and backward directions with reference to the initial direction of muon beam.

The advantages of measuring longitudinal relaxation are manifold. First, $G_z(t)$ reflects both the static and dynamic character of the local field. Second, longitudinal relaxation can be observed with and without the external field; one of the unique features of the muon probe is that we can measure $G_z(t)$ at zero external field (ZF-μSR). With the help of theoretical developments initiated by Kubo and Toyabe (1967), the motion of the spin direction, seen in the time-evolution spectrum of the anisotropic e^+/e^- decay, can be given a one-to-one correspondence to the static or dynamic nature of the microscopic magnetic field seen by the muon (Figure 6.5).

The one-to-one correspondence, like a ZF-μSR guide map, is classified into four cases according to the nature of the local field, in both static and dynamic instances: (1) a unique

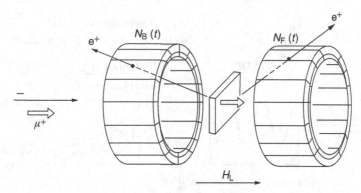

Figure 6.4 Typical layout for a muon spin rotation/relaxation/resonance (μSR) experiment with pulsed muons. Geometries for longitudinal relaxation with the field applied parallel to the beam direction with segmented counter telescopes.

and uniform local field like internal field (H_{hf}) of a magnetically ordered system exhibiting a spin rotation around the local field vector (note that $<\cos^2\theta> \neq <\sin^2\theta>$ is assumed), where, in the dynamic case, three extreme cases are considered; (2) a unique local field with some inhomogeneous broadening appearing in many cases in μSR exeriments on newly developed crystals where field inhomogenities come from either crystal imperfections or impurities; (3) random and directionally isotropic local field with a Gaussian distribution in field strength, appearing in nuclear dipolar field at the interstitial μ^+ site, where, in the dynamic case, a motional narrowing effect is considered; (4) random and directionally isotropic local field with a Lorentzian distribution in field strength, appearing in a local field distribution at μ^+ site in a dilute-alloy spin-glass system where, in the dynamic case, a motional narrowing effect is considered.

Details of some typical cases presented in Figure 6.5 of ZF-μSR as well as of LF-μSR (muon spin relaxation under longitudinal field) are given below.

6.2.1 Some details of zero-field relaxation functions

Derivation of the relaxation functions are given for some typical cases shown in Figure 6.5. More detailed explanations are given in the paper by Hayano *et al.* (1979).

ZF relaxation under static random fields

Assuming a Gaussian distribution of isotropic random fields such that $p(H_i) = N_p \exp[-(\gamma_\mu^2 H_i^2)/(2\Delta^2)]$, where $N_p = (\gamma_\mu)/(\sqrt{2\pi} \cdot \Delta)$, where $i = x, y, z$, and Δ^2/γ_μ^2 is the second moment of the random field along all three directions, we have, after statistical averaging, the following result, as seen in Figure 6.5:

$$G_z(t) = \frac{1}{3} + \frac{2}{3}(1 - \Delta^2 t^2) \exp\left(-\frac{1}{2}\Delta^2 t^2\right)$$

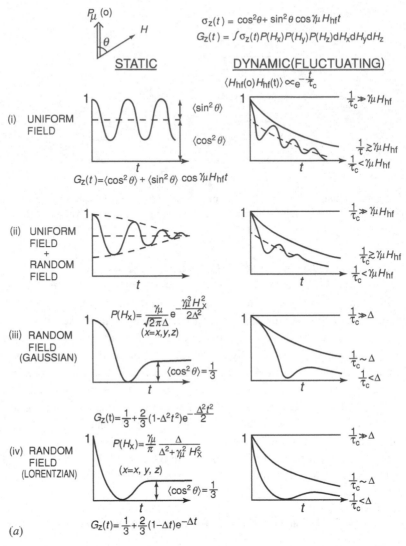

Figure 6.5 (a) The zero external field muon spin rotation/relaxation/resonance (ZF-μSR) time-spectra expected from microscopic local fields experienced by the muon with various magnitudes, distributions, and characteristic time constants of dynamic fluctuation (correlation times). Details are given in the text.

Similarly, assuming a Lorentzian distribution of random fields such that $p(H_i)(\gamma_\mu/\pi) \cdot \Delta/(\Delta^2 + \gamma_\mu^2 H_i^2)$, we have the following relaxation function:

$$G_z(t) = \frac{1}{3} + \frac{2}{3}(1 - \Delta t)e^{-\Delta t}$$

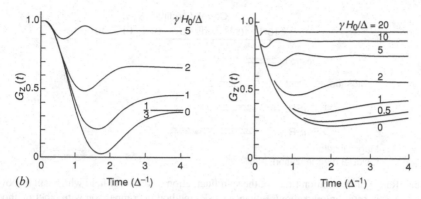

Figure 6.5 (*b*) The muon spin relaxation under longitudinal field (LF-μSR) time-spectra with applied longitudinal field for the μ^+ spin under random field of static Gaussian distribution (left) and Lorentzian distribution (right).

ZF relaxation in the presence of dynamical effects

When the muon is in the presence of fluctuating fields with a mean frequency v, the strong collision model (under the assumption that there is no correlation between the fields before and after the change) can give the relaxation function $G_z(t)$ until time t with a number of changes n, by using static relaxation function for the instantaneouly static field as follows (Hayano *et al.*, 1979):

$$G_z(t) = \sum_{n=0}^{\infty} p^{(n)}(t)$$

$$p^{(0)}(t) = e^{-vt} G_z^{st}(t)$$

$$p^{(1)}(t) = v \int_0^{\infty} dt_1\, e^{-v(t-t_1)} G_z^{st}(t - t_1) \times e^{-vt_1} G_z^{st}(t_1), \text{ and so forth.}$$

With the help of Laplace transformation, the following formula can be obtained:

$$G_z(t) = \int_0^{\infty} f(s) e^{st} \alpha t$$

$$f(s) = \frac{1}{3s} + \frac{2}{3} \frac{s}{\Delta^2} \left[1 - s \int_0^{\infty} \exp\left(-\frac{1}{2}\Delta^2 t^2 - st \right) dt \right]$$

For slow fluctuations, the following expression, in which there is a suppression of the $\frac{1}{3}$ term leading to a hump at $t \sim 3/\Delta$, is obtained:

$$G_z(t) \sim \frac{1}{3} \exp\left(-\frac{2}{3} vt \right), (t >> 3/\Delta)$$

Thus, not only the transverse relaxation rate (λ_T) seen in the muon spin rotation experiment, but also the ZF longitudinal relaxation rate (λ_L) can be related to the fluctuation frequency (v) or the correlation time ($\tau_c = v^{-1}$). The observed dynamic behavior of the surrounding magnetism which can be probed by the μSR is compared to the cases seen in other experimental methods in terms of the range of correlation time, as shown in Figure 6.6, where the

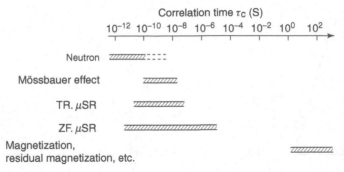

Figure 6.6 Range of correlation times of the spin fluctuation of the local field which can be covered by the muon spin rotation/relaxation/resonance (μSR) method in comparison with other methods, where the fluctuation of a 1 μ_B spin moment at a distance of 1 Å(0.1 nm) from the muon is assumed.

fluctuation of a 1 μ_B spin moment at a distance of 1 Å(0.1 nm) from the muon is considered. The unique feature of the μSR method is clearly seen.

6.2.2 Spin relaxation under longitudinal field: LF-μSR

Once the relaxation is seen in the ZF-μSR spectrum, the nature of relaxation, where the muon is under static or dynamic field, can be distinguished by applying the longitudinal field.

Static case: LF-decoupling of random static field

ZF-μSR, seen in Figure 6.5(*a*) (cases (ii) and (iv)) can be decoupled: the original muon spin polarization can be restored by applying a sufficiently strong applied field.

In the case of an isotropic Gaussian random field, the effect of longitudinal field B_0 ($\omega_0 = \gamma_\mu B_0$) can be written as follows (Kubo and Toyabe, 1967; Yamazaki, 1979):

$$G_z(t) = 1 - \frac{2\Delta^2}{\omega_0^2}\left[1 - \exp\left(-\frac{1}{2}\Delta^2 t^2\right)\cos\omega_0 t\right]$$
$$+ \frac{2\Delta^4}{\omega_0^3}\left[\int_0^t \exp\left(-\frac{1}{2}\Delta^2\tau^2\right)\cos\omega_0 t\, d\tau\right]$$

The $G_z(t)$ is shown in Figure 6.5(*b*).

Similarly in the case of the Lorentzian random field, the result can be obtained as shown in Figure 6.5(*b*) (Kubo, 1981; Uemura, 1981).

Dynamic case: LF-decoupling of fluctuating field

When muon spin is subject to a strong and rapid fluctuating field $\bar{\mathbf{H}}(\bar{H}_x, \bar{H}_y, \bar{H}_z)$, the spin relaxation phenomenon under an applied longitudinal field H_0 is helpful in estimating a correlation time (τ_c). This parameter characterizes the spin dynamics of the fluctuating field: $H_i(t)\bar{H}_i(t+\tau) = \bar{H}_i^2 \, e^{-\tau/\tau_c}$ ($i = x, y, z$).

Following the Redfield theory frequently used in various magnetic resonance studies (Slichter, 1965), the spin relaxation rate parallel to longitudinal field $1/T_1$ and that perpendicular to longitudinal field, $1/T_2$ can be written as follows:

$$1/T_1 = \gamma_\mu^2 [\bar{H}_x^2 + \bar{H}_y^2] \frac{\tau_c}{1+\omega_0^2\tau_c^2}$$

$$1/T_2 = \gamma_\mu^2 \left[\bar{H}_x^2 \tau_c + \bar{H}_y^2 \frac{\tau_c}{1+\omega_0^2\tau_c^2} \right]$$

where ω_0 is the Larmor frequency of the applied longitudinal field; $\gamma_\mu H_0 = \omega_0$.

Thus, $1/T_1$ becomes maximum at around $\tau_c = 1/\omega_0$ so that a measure of τ_c can be obtained by the longitudinal field dependence of the relaxation rate.

6.2.3 Longitudinal field decoupling of muonium (Mu)

From the very beginning of muonium studies, the LF decoupling experiment has been employed to observe muonium signals in various kinds of condensed matter by assuming absence of spin conversion and chemical reaction:

$$A = A_{Mu} \frac{(1 + 2x^2)}{2(1 + x^2)}$$

where A_{Mu} is the initial asymmetry of μ^+ in Mu and x is the ratio of the applied field to the Mu hyperfine field (1585 G in the case of free Mu in vacuum). The LF decoupling pattern can be obtained by measuring the time-averaged muon polarization against the applied longitudinal field. The quantity $A_{Mu}(x = 0)$, which is equal to 1/2 in the absence of perturbations, varies from 1/2 if Mu experiences random perturbing fields from the surrounding nuclear dipoles; if the fields are static it takes the value 1/6, while if the fields fluctuate dynamically it becomes zero.

The time-evolution of the forward/backward asymmetry $AG_z(t)$ of the decay e^+ from the μ^+ in the Mu which is subject to spin-flip conversion ($I = 1 \rightarrow I = 0$) and chemical reaction ($M \rightarrow \mu^+$) can be described using the depolarization model developed by the Russian group (Ivanter and Smilga, 1969). When both the reaction rate ($1/\tau$) and the spin-flip probability (ν) are much smaller than the Mu hyperfine frequency $2\pi\nu_0$ ($\nu_0 = 4.463 \times 10^9 \, s^{-1}$ in vacuum), $A_D(t)$ changes with time as:

$$AG_z(t) = A_{Mu} \frac{1 + 2x^2}{2(1 + x^2 + \nu\tau)} \times \left(1 - \exp\left[-\left(\frac{1}{\tau} + \frac{\nu}{1 + x^2}\right)t\right]\right)$$

The normalized time-integrated asymmetry, which describes a change in asymmetry against external field appearing in the decoupling pattern, is expressed by:

$$A\bar{G}_z = 1 - \frac{(2\pi\nu_0 r)^2(1 + 2\nu_c\tau)}{2[(2\pi\nu_0\tau)^2(1 + \nu_c\tau + x^2) + (1 + 2\nu_c\tau)^2]}$$

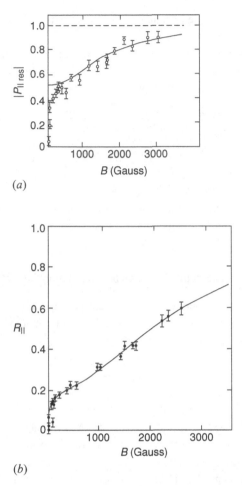

Figure 6.7 Example of decoupling pattern of the residual average muon polarization against applied longitudinal field fitted using appropriate values of the hyperfine field (v_0), spin conversion rate (τ), and chemical reaction rate (v_c). (a) Al_2O_3 (Minaichev *et al.*, 1970) and (b) KCl (Ivanter *et al.*, 1972).

where v_0 is the hyperfine frequency of the muonium-like state, and v_c and $1/\tau$ are the spin conversion rate and the chemical reaction rate of muonium, respectively. This formula has been applied to explain a strange field dependence of the asymmetry observed for the μ^+ in alkali halides (Ivanter *et al.*, 1972). The experimental method has also been utilized in order to understand the origin of the "missing" fraction of μ^+ in various molecular liquids.

As expected from the form of the above formula, the LF decoupling pattern cannot be interpreted uniquely; various sets of v_0, τ, and v_c can give us the same shape of $A_{av}(x)$ (Figure 6.7). Various new spectroscopic methods have been developed for the study of Mu under longitudinal applied field, and these have been applied in recent years to yield quantitative basic information regarding Mu in solids, as described later.

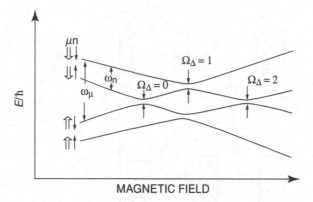

Figure 6.8 Conceptual picture of the principle of level-crossing resonance presented in the form of the energy diagram for a muon and a single-spin-1/2 nucleus coupled by some interactions. The energy levels tend to avoid one another near resonance by an amount of Ω_0 (Kiefl and Kreitzman, 1992).

6.2.4 Level-crossing resonance (LCR)

In muonium or muonium-substituted radicals, the μ^+ with the bound electron is subject to hyperfine fields from the surrounding localized nuclei with various hyperfine coupling constants. When the μ^+ Zeeman frequency corresponding to the applied longitudinal field becomes equal to one of these hyperfine splittings, there is a strong energy transfer between the μ^+ and the electron-bath system, causing a substantial reduction in the μ^+ longitudinal polarization.

As the simplest example, let us consider muonium with a hyperfine coupling constant A_μ, in the vicinity of the single-spin-1/2 nucleus which is subject to the hyperfine field A_n (Figure 6.8). The two mixing states appear to avoid one another on resonance as a result of their interaction and thus do not cross in energy, so that the name of avoided-level-crossing (ALC) is frequently used in place of LCR. The resonance signal can be seen as an increase in muon spin depolarization, since the spin oscillates with the frequency of the hyperfine coupling constant. By solving a 2×2 matrix representing spin Hamiltonian, the LCR occurs approximately at the field H_R, expressed by the following formula:

$$H_R = \frac{|A_\mu - A_n|}{2(g_\mu \mu_\mu - g_n \mu_n)}$$

Thus, the LCR method can provide us with a precise value for the μ^+ hyperfine field even under a decoupling field.

The LCR can occur in different types of energy transfer among two subsystems, where one subsystem involves the muon spin. In Cu, where the first LCR was observed, the energy matching occurs between Zeeman splitting of the μ^+ spin and nuclear quadrupole splitting of the nearby Cu nuclear spin induced by the presence of the μ^+.

The phenomenon of LCR was first proposed by Abragam (1984), then experimentally realized by the μSR group at Tri-University Meson Facility (TRIUMF) for μ^+ in Cu and

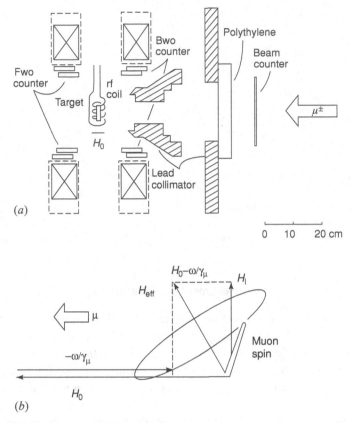

Figure 6.9 (*a*) Schematic view of the experimental arrangement for muon spin resonance experiments, including counters, r.f. coil, and applied fields. (*b*) Conceptual view of magnetic resonance with r.f. field represented in the rotating frame coordinate system.

for Mu radicals in tetramethylethene (Kiefl *et al.*,1986; Kreitzman *et al.*, 1986a; Kiefl and Kreitzman, 1992).

The width of LCR is determined by the splitting on resonance between the two mixed levels corrected by the muon natural width. Line shape of the LCR signal is affected by spin relaxation to provide important information regarding, e.g., μ^+ diffusion in a crystal (Kreitzman *et al.*, 1986a), electron-spin relaxation and a chemical reaction in muonium-substituted radical (Heming *et al.*, 1989).

6.3 Muon spin resonance

Exactly analogously with conventional nuclear magnetic resonance, we can induce magnetic resonance of a muon spin placed under a local magnetic field (Kitaoka *et al.*, 1982; Nishiyama, 1992). A typical experimental arrangement is shown in Figure 6.1. The principle of magnetic resonance is also illustrated in Figure 6.9. First, let us apply a longitudinal field H_0 which creates a Zeeman splitting. This does not change the initial muon spin direction. Then, we apply an r.f. field perpendicular to the muon spin direction so as to induce emission and absorption of photons of frequency ω. When ω matches the Zeeman frequency ω_0 due to the local field H_{loc} ($\omega_0 = \gamma_\mu H_{\lambda_{0\chi}}$), transitions among the Zeeman levels take place. This resonance can be detected via changes in the asymmetric distribution of decay electrons. The quantity to be observed is:

$$N(\theta, t) = N_0 \exp(-t/\tau)[1 + AG_z(t, \omega) \cos\theta]$$

in which $G_z(t, \omega)$ stands for the attenuation factor as a function of ω.

How $G_z(t\omega)$ behaves can be seen from an intuitive picture of magnetic resonance (Figure 6.9). In a rotating frame of frequency ω, the r.f. field, $2H_1\cos\omega t = H_1[\exp(i\omega t) + \exp(-i\omega t)]$, is static and the effective field that the muon experiences in this rotating frame is:

$$\mathbf{H_{\text{eff}}} = \mathbf{H_0} + \omega/\gamma_\mu + \mathbf{H_1}$$

So long as ω is far from $-\gamma_\mu H_0$ (off-resonance), H_{eff} is along the direction of H_0, but in the region $\omega = -\gamma_\mu H_0$ (near-resonance), the r.f. field H_1 plays an important role, as the H_{eff} turns towards the direction perpendicular to H_0. In this region, the spin rotates around the x-axis, and the longitudinal component of the spin (or time-differential attenuation factor) is described as:

$$\sigma_z(t) = \cos^2\beta + \sin^2\beta \, \cos\gamma_\mu H_{\text{eff}}t = G_z(t, \omega)$$

where:

$$\tan\beta = H_1/(H_1 + \omega/\gamma_\mu)$$

Let us suppose that the muon state undergoes a transition from an initial state to a second state where the muon feels a local field of H_i and H_f, respectively. In magnetic resonance one can detect both signals as two resonance frequencies, $\omega_i = \gamma_\mu H_i$ and $\omega_f = \gamma_\mu H_f$, whose amplitudes change with time. By contrast, transverse field–muon spin rotation (TF-μSR) cannot exhibit such information, because, for the precession to be detected, the phase has to be preserved. The first field may be observed as a precession pattern, but it damps as $\exp(-\lambda t)$ with a transition time constant λ. The second precession loses its phase relationship with the first one in a short time $t_{\text{dephase}} \cong 1/(\omega_1 - \omega_2)$. Except for some special cases, $\omega_1 \cong \omega_2$ or λ larger than t_{dephase}, there is no way to observe the second state by TF-μSR. A number of physical situations lead to this kind of dephasing: muonium $\to$ diamagnetic muon in chemical reactions, diffusion$\to$trapping$\to$detrapping in some diffusion processes. In such cases the magnetic resonance method plays a unique role, and in fact, in some materials, hitherto unknown states of μ^+ behavior have been revealed by the resonance method.

6.4 μ^+SR, MuSR, and μ^-SR

As summarized in Figure 6.2, there are three types of μSR method depending upon the character of the muon probe inside the condensed matter to be probed. The following sections outline these methods in a little more detail.

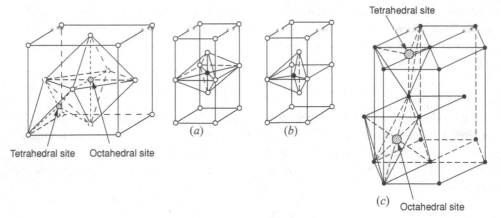

Figure 6.10 Typical interstitial sites of representative crystal structures: (*a*) face-centered cubic or fcc; (*b*) body-centered cubic or bcc; and (*c*) hexagonal close-packed or hcp.

6.4.1 μSR of diamagnetic μ^+ : μ^+SR

Diamagnetic μ^+ is the μ^+ state typically formed when energetic μ^+ is introduced into metals and other high-electron-density materials. It maintains its full initial polarization and emits e^+ with a characteristic spatial distribution (see Chapter 1).

In principle, the interstitial location of the μ^+ can be determined from the electrostatic potential at the interstitial sites with a correction for dilation of the surrounding lattice. Typical interstitial sites in representative crystal structures are shown in Figure 6.10.

Diamagnetic μ^+ can also be formed if the muon bonds chemically to, for example, a negatively charged O or N site, e.g. in complex compounds such as high-T_c superconductors (Sulaiman *et al.*, 1994). The determination of the μ^+ site and the type of bonding are central subjects in μ^+SR, and are covered further in Chapter 8. In some magnetic materials, there may be a muonium-like state whose spin, coupled antiparallel to the magnetic spin, behaves like a diamagnetic state. Such a μ^+ state is found in organic ferromagnets (Jeong *et al.*, 2002).

In condensed-matter applications of this kind diamagnetic μ^+ can function either as a passive probe of the material's intrinsic microscopic magnetism or as an active probe of a new microscopic environment; the change is induced by the presence of the μ^+, including effects such as the diffusion of the μ^+ itself.

6.4.2 μSR of paramagnetic Mu: MuSR

Once a stable paramagnetic Mu state is formed in condensed matter, the most conclusive evidence of this formation is a characteristic response to an applied magnetic field; Mu exhibits a rotation frequency 103 times more rapid than diamagnetic μ^+. Another piece of evidence that Mu formation has occurred is a characteristic increase in asymmetry against the increase of applied longitudinal field; in other words, the existence of a decoupling pattern (see section 6.2.3).

The location of neutral Mu can be predicted by carrying out energy minimization, with correct treatment of the response of the surrounding lattice atoms to the presence of the muonium. Experimentally, the location and electronic structure of Mu and related paramagnetic μ^+ states in condensed matter can be determined by Mu spin rotation, LCR, and other techniques.

The main roles of MuSR in condensed-matter studies are to probe the nature of hydrogen-like centers and defects, including those related to isotopic mass dependence, and to investigate the chemical reactions of Mu (as an H analog) in solids and liquids.

6.4.3 μSR of bound μ^-: μ^-SR

The injection of energetic μ^- into condensed matter leads, without exception, to the formation of a muonic atom and the cascading-down to the ground state of the muonic atom, where the μ^- passes most of its lifetime. The nature of the bound μ^- in terms of lifetime and magnetic moment is described generally in section 4.1.

The characteristic signal of the bound μ^- in various elements is a unique lifetime which reveals information about the nucleus to which the μ^- is bound. In the sense of condensed-matter applications, the nature of μ^- bound to some nucleus of atomic number Z is a dilute impurity with charge state of $(Z - 1)$ which behaves as a spatially expanded nucleus (Figure 6.2). At a specific site in some ionic material, such as an apical oxygen in a high-T_c superconductor, the bound μ^- is surrounded by electrons with one hole, producing a unique paramagnetic hole probe.

The role of μ^-SR is to probe directly the local character of the magnetic hyperfine interaction adjacent to the nucleus. The implantation of μ^- into matter can be considered as equivalent to the formation of an acceptor center (in semiconductors: μ^- functions as an Al center in Si) or as the introduction of spin to a spin-zero nucleus (with reference to the corresponding NMR studies).

6.5 Experimental methods of μSR: continuous vs pulsed

As described in Chapter 2 (section 2.1.1), there are two types of muon beams available for μSR studies: continuous, from cyclotron (Paul Scherrer Institute (PSI), Tri-University Meson Facility (TRIUMF), etc.) and pulsed, from rapid-cycling synchrotron (Rutherford Appleton Laboratory (RAL), High Energy Accelerator Research Organization (KEK), etc.). Among the features of each of these two beams, the following ones are the most practically important in various types of μSR experiments.

In continuous μSR time resolution is higher, while it is limited to $2\Delta t_w$ (Δt_w: time width of the beam pulse) in pulsed μSR, e.g. 100 ns (10 MHz) for 50 ns pulse-width. In continuous μSR the time-range for the μSR time-spectrum is limited by the next-incoming muon so that, without reducing muon intensity, the available time range is limited; a few μs for 10^6/s stopping muon, while in pulsed μSR the μSR time-range is longer (up to 20 μs or more) in a rate-unlimited manner. Coupling with strong r.f. and/or laser is easier in pulsed μSR so that various types of resonance experiments can be made.

All μSR experiments involve the following steps:

1. Stopping muons in a target sample after passing through a certain thickness of energy degrader. Because of the momentum spread, the stopping region extends over some finite range, typically (as estimated from the arguments given in Chapter 3) 5 g/cm^2 for 200 MeV/c muons, 2 g/cm^2 for 100 MeV/c muons, and 100 mg/cm^2 for 4 MeV surface muons. For ultraslow μ^+ of below 10 keV, a thickness can be as low as sub μm.
2. Focusing and/or collimating the muon beam into the target with the removal of beam halo and contamination as completely as possible. A typical beam size at the target is 2×2 cm or smaller. With a continuous beam, one can define the beam region using a counter logic involving the shape of the defining counter.
3. Identification of the stopping muons (continuous μSR only).
4. Detection of the decay positron/electron using a counter system.
5. Precise time measurement of the interval between the stopping muon and detection of the decay electrons.
6. Data acquisition and processing.

The experimental techniques have important variations depending on the time character of the muon beam, that is, whether it is continuous or pulsed.

6.5.1 Continuous μSR

Typical examples of beam/target/counter arrangements for μSR in a continuous beam ("continuous μSR") experiment are shown in Figure 6.11. Each of the incoming muons is identified by a series of plastic scintillation counters: B = beam counter, M = muon timing counter, D = beam defining counter, and X = veto counter against straight-through muon events. The D counter has to be adjusted so as to cover only the target size to be selected. It must be thin, because muons stopping here cannot be discriminated from those stopping in the target – below 0.1 mm for 4 MeV surface μ^+. The veto counter X covers a large enough area to detect any directly transmitted muons.

Then, we define the following logic: "stopped μ" $= B \times M \times D \times \bar{X}$. Decay electron events produce signals in the corresponding counter telescopes: $E1, E2$ = electron counters. Again, we define the appropriate logic: "decay e" $= E1 \times E2 \times (\bar{B} + \bar{M})$. There, symbols $\times, +$ and $^-$ represent "and" (coincidence), "or," and "veto" (anticoincidence), respectively.

When the beam has a microscopic pulse structure due to r.f., the contamination of electrons has the same r.f. structure, and thus the accidental background exhibits regular spikes with the r.f. period.

The next stage in the electronics is typically a time-to-digital converter (TDC), where the time interval measurement is usually initiated by a stopped-μ event and terminated by a decay-electron event. Usually, a time digitizer with a high time resolution (better than 1 ns) and a wide time range (up to 10 μs or longer) is used.

When two muons arrive successively before an electron event, a problem arises about which muon has time correlation with the electron. If the electron is the decay product of the second muon, the time spectrum started by the first muon constitutes background which

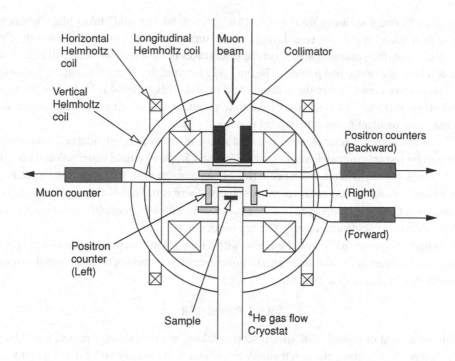

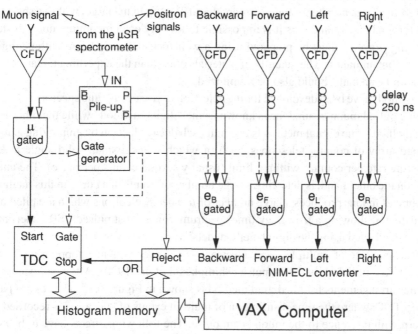

Figure 6.11 Example of experimental and electronics/computer arrangement for continuous muon spin rotation/relaxation/resonance (μSR). CFD, constant fraction discriminator; TDC, time-to-digital converter (Kojima, 1995).

increases as one goes away from $t = 0$. This "growth background" takes place whenever more than one start-pulse occurs during the measuring time interval, and distorts the time spectrum from the genuine shape. A similar effect takes place when two successive electrons are detected after a stopped μ event. The only way to avoid this type of distortion is to reject any successive events within the measurement period. This "second μ" rejection limits the total stopped μ rate to the inverse of the time range. In this continuous mode one cannot increase the incident muon rate beyond this limit.

Since the logic for both stopped μ and decay e involves anticoincidence, those events covered by the anticoincidence gates are killed, and we have a time distribution with a dip in the $t = 0$ region. When a muon beam free of contamination stops fully in the target, there is no need for such anticoincidence vetoes, and so there is no loss of the $t = 0$ region. This arrangement is always employed in μSR experiments with a high-quality surface μ^+ beam, where the anticoincidence counter X is not needed at all.

For high-frequency μSR measurements, with a time resolution of 0.1 ns or better, special care must be taken not to introduce an additional time distribution of e^+ hitting on the timing counter due to spatial distributions.

6.5.2 Pulsed μSR

With the advent of pulsed μSR spectroscopy making use of a sharply pulsed muon beam (Nagamine, 1981), new detection methods have been continuously under development. In a pulsed beam, a large number of muons (more than 10^3) stop in the target within a short time interval (around 50 ns), and thus it is impossible to identify each incoming muon. Instead, the stopping μ events must be prepared using a well-focused muon beam with negligibly small e/π contamination. The quantity of materials other than the experimental target lying in the muon beam path should also be minimized.

Two methods have been developed for the detection of μ-e decay time spectra in conjunction with a pulsed muon beam. One is known as the "digital method" while the other is the "analog method." The digital method is to count each decay electron by employing a highly segmented array of counter telescopes, each of which spans a small solid angle to keep the counting rate per counter within a limit (usually several events per pulse). The analog method, on the other hand, is detecting a large number of events in a burst; in this method, a single burst of muons produces a large number of μ-e decay electrons which are piled up in a large detector, showing a μ-e decay time spectrum. Since most pulsed μSR experiments have been conducted using the digital method, details are only given here for that technique.

A pulse sequence of multiple decay electrons is recorded in a nonstopping mode with reference to the initial muon beam pulse (Shimokoshi et al., 1990). A typical example of electronics arrangements for the digital method is shown in Figure 6.12. The central part is a nonstop TDC, where the time distribution of multihit events of "decay e" is recorded as a bit pattern with reference to the muon beam pulse. The following effects have to be taken into account:

1. Counting-loss effect in the nonstop TDC. Practically, the TDC has a finite time bin. Therefore, when more than two signals fall in the same time bin they are accepted as one

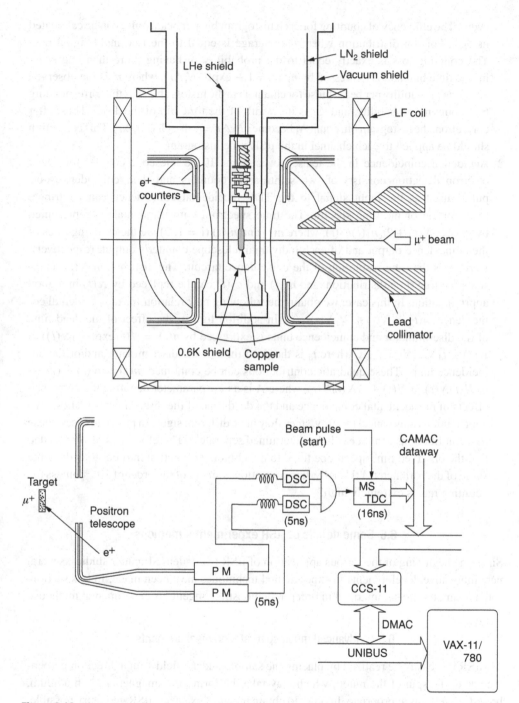

Figure 6.12 Example of experimental and electronics/computer arrangement for pulsed muon spin rotation/relaxation/resonance (μSR). CCS-11, microprogrammed CAMAC processor, DMAC, direct memory access; MS TDC, multistop time-to-digital converter; PM, photomultiplier (Kadono *et al.*, 1989).

event. The efficiency of counting for each histogram bin per beam pulse can be calculated using a Poisson distribution where the average is equal to the true input signal rate. The counting loss is exactly equal to the probability of having more than one event in one time bin, and is expressed by $n_i \Delta t = 1 - \exp(-N_i \Delta t)$, where n_i is the observed counting probability per beam pulse for channel i in the histogram, N_i is the corresponding true counting probability, and Δt is the width of the time bin of the TDC. Thus, after correction, the histogram bin count N_i becomes $N_i \Delta t = -\log(1 - n_i \Delta t)$. This correction should be applied to each channel in the μSR time histogram.

2. Accidental coincidence in the telescope counters. If the telescope counter for decay electron detection consists of two scintillation counters and their coincidence output is used as the input signal to the TDC, accidental coincidence can contribute to distortion of the time spectra. The time spectrum with accidentals is represented by $n(t) = N(t) + 2t_c n_1(t) n_2(t)$, where $n(t)$ and $n_i(t)$ (i = 1, 2) are the counting rates of the coincidence output and of each individual telescope counter's output, respectively. Here, t_c is the time resolution of the coincidence circuit. The distribution $N(t)$ represents the true time distribution, and $n_1(t)$ and $n_2(t)$ can be replaced by $N(t)$ in a good approximation. In this case, we obtain the following new relation for the accidental coincidence: $n(t) = N(t) + 2t_c N(t) N(t)$. In addition to this, the effect of the dead time of the discriminator and coincidence units is expressed by $n(t) = N(t) \exp(-t_d N(t)) = N(t) - t_d N(t) N(t) + \ldots$ where t_d is the dead time of the discriminator and/or the co-incidence units. These quadratic contributions can be combined: $n(t) = N(t) + (2t_e - t_d) N(t) N(t) = N(t) + A N(t) N(t)$, where A is a new parameter denoting the combined effects of the accidental coincidence and the dead time of the discriminators. These two contributions may cancel roughly, since they have different signs. In practical cases, these two contributions cannot easily be determined separately. Therefore, it is recommended that the observed time spectra be fitted to the combined function in order to deduce the value of the parameter A. Usually, this correction is small, of the order of 10^{-3}, unless the counting rate is extremely high.

6.6 Some details of μSR experimental methods

Since the beginning of the serious application of μSR to condensed-matter studies, several new innovative developments in experimental techniques have been made. In this section, some examples are summarized in order to demonstrate ingenious experimental methods.

6.6.1 Advanced muon spin rotation measurements

TF-μSR can easily be realized by placing the sample under a field with a direction perpendicular to the spin of the muon, which may take the form of diamagnetic μ^+, muonium, bound μ^-, etc., as appropriate. In order to obtain the most exact TF-μSR geometry possible, several precautions are required with respect to the experimental arrangements.

As mentioned earlier, there are two types of TF-μSR geometry: (1) inject longitudi-nally spin-polarized muons into a sample placed in a transverse magnetic field (normal geometry); (2) inject transversely spin-polarized muons into a sample placed under longitudinal field (rotated geometry). In the latter case, the spin rotator must be installed with a careful adjustment to yield the true 90° spin direction.

In regard to low-field TF-μSR, stray fields must be eliminated using correction coils. This is particularly important in the case of Mu spin rotation; in a 1 G Mu-spin rotation experiment, the stray field must be kept below 10 mG.

As for high-field TF-μSR, if the aim is high-precision frequency determination, in addition to the problems due to deflection of the incoming μ^+ trajectory, it is necessary to consider the distribution of the e^+ stopping region; these two effects may lead to finite timing distributions respectively of the μ and the e. In order to minimize these phenomena, spin-rotation geometry should be employed. An advanced high time-resolution spectrometer for the μ^+ spin rotation under field up to 4 T has been developed at TRIUMF, as shown in Figure 6.13.

6.6.2 Advanced longitudinal relaxation measurements

Under zero field

The longitudinal relaxation of the muon spin $G_z(t)$ can be determined by measuring $N(\theta, t)$ in the forward and backward directions:

$$\frac{N(\theta, t)}{N(180, t)} = \alpha \frac{1 + AG_z(t)}{1 - AG_z(t)}$$

allowing some instrumental asymmetry α. The existence of this instrumental asymmetry parameter is the result of several factors, including the following:

1. The target sample is not located exactly centrally between the two counters.
2. The stopping region of the muon is not at the center of the target thickness.
3. There may be variations in solid angle efficiency from counter to counter.

In order to minimize the first contribution it is recommended that a special device be prepared for fine adjustment of the target location.

Even after minimizing the instrumental asymmetry, it is necessary to determine the value of the effective asymmetry parameter α. For this purpose, it is convenient to install a transverse-field source as a supplement to the ZF- and LF-μSR apparatus. The TF spin rotation signal provides a baseline for the time spectrum of the forward/backward ratio.

Another method to determine α is the use of "internal calibration," i.e., self-determination of α from the forward/backward ratio at a time much longer than the relaxation time. This method can be applied to magnetic materials. This kind of internal calibration is particularly effective in the presence of background contributions from materials other than the target, e.g., cryostat wall.

For true ZF-μSR, in particular for Mu, it is essential to establish an accurate ZF setting corrected for stray magnetic fields from the surrounding magnets, the earth's magnetic field, and any other external sources. The tolerance on this setting should be smaller than

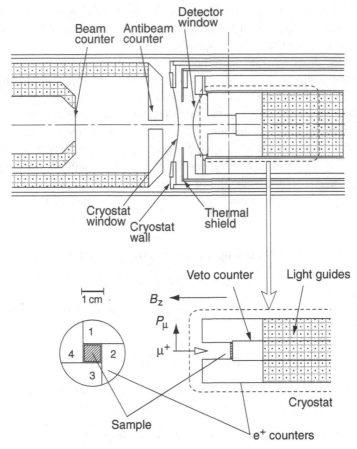

Figure 6.13 Example of high time-resolution TF-μSR spectrometer developed at Tri-University Meson Facility (TRIUMF). Transversaly polarized μ^+ is injected into the specimen placed under 4 T longitudinal field. Veto connector is used to reject the μ^+ passing through the specimen. A whole system is inserted into a warm bore of 4 T superconducting coil.

10 mG. For this purpose, correction field coils in the x-, y-, and z-directions should be installed.

Under external longitudinal field

The experimental arrangement for LF-μSR is essentially the same as for ZF-μSR except for the presence of an additional Helmholtz coil pair to produce the longitudinal field. This method is often used to measure the field dependence of $G_z(t)$. However, we must be careful about any field-dependent effects in the experimental arrangements which may affect the asymmetry. For example:

1. A change in the μ trajectory and thus the stopping position along the target thickness causes a change in the "baseline."

2. A change in the decay e trajectory as a function of longitudinal field strength causes a field dependence in the detector solid angle.
3. When the low-energy component of the decay electrons is spatially confined due to the magnetic field, the asymmetry increases with the field.
4. The energy loss of the decay electron inside the target also has a field dependence.

All of these effects should be checked by using materials which have known properties under LF-μSR. For instance, μ^+ in pure Al at room temperature always yields a field-independent full asymmetry, while μ^+ in some rare-earth ferromagnets (Ho, Er) has relatively fast paramagnetic relaxation (of the order of μs^{-1}), which is almost field-independent up to 4 T.

Another powerful method of baseline determination under applied longitudinal field is to apply the muon spin resonance method. One can reverse the direction of the muon spin by applying a $180°$ pulse of r.f. field without changing any of the experimental conditions.

The four effects mentioned above can be predicted by Monte Carlo simulation for a given field and counter geometry.

6.6.3 Advanced muon spin resonance measurements

In muon spin resonance experiments one applies a longitudinal field H_0 and a transverse r.f. field $2H_1\cos\omega t$ to measure the attenuation factor $G_z(\omega)$, in the forward/backward asymmetry. In order to induce resonance, that is to say, spin rotation with respect to H_1, within the short lifetime of the muon, we need a large r.f. field amplitude. The peak power P_{rf} and the strength of H_1 are related through the expression $P_{rf} = C \times H_1^2 \times V$. For H_1 of 40 G with a coil volume of 10 cm^3, the required peak power is 40 kW. This is nearly impossible if the r.f. field is continuously applied. Therefore, the use of pulsed muon with a reasonably low-duty factor operation is quite reasonable to be combined with a spin resonance experiment with a pulsed r.f. field.

For higher r.f. frequencies, the r.f. field should be generated in a cavity. For an r.f. field generated in the TM$_{110}$ mode, the diameter of the cavity (d) is related to the frequency by $d(\text{cm}) = 0.146 \times f$ (MHz). For 500 MHz muon spin resonance, the diameter of the cavity is 73.2 cm, which should be installed inside a superconducting longitudinal field of 37 kG (to match the resonance field of free μ^+). With such a set-up, diamagnetic muon states can be probed under around 40 kG longitudinal field, while free muonium can be observed under a field of 400 G (ν_{12} resonance).

In the r.f. resonance experiment, both the size and the form of the sample should be chosen with consideration of the r.f. skin depth d_s, with $d_s(\text{cm}) = \alpha[f(H_z)\kappa_c]^{-0.5}$ where κ_c is the conductivity of the sample in units of (ohm)$^{-1}$. Typically, a metallic sample may be either powdered or a stack of thin foils with insulating interlayers.

The most significant advantage of the time-differential resonance technique is the capability to detect the μ^+ state after a state change leading to the disappearance of the initial phase coherence. As a typical example, the time evolution of a diamagnetic state formed after the reaction of a precursor muonium (muonic radical) state is formulated below. At resonance, the transition between the two spin substates of the diamagnetic μ^+ state is observed in a time-differential fashion through the changing forward/backward asymmetry of the decay e^+, which exhibits a spin rotation pattern around the applied RF field H_1. The asymmetry $AG_z(t)$ of the diamagnetic μ^+ at resonance is the sum of the asymmetry of the "prompt" diamagnetic μ^+ species formed at time zero and that of the time-delayed diamagnetic μ^+ species produced after the Mu reaction. The time evolution of $AG_z(t)$ is given by:

$$AG_z(t) = A_\mu e^{-\alpha t} \cos 2\pi \nu_1 t + \int_0^t \frac{dA_D(t')}{dt} dt' e^{-\alpha(t-t')} \times \cos 2\pi \nu_1 (t - t')$$

where A_μ is the initial asymmetry of the prompt diamagnetic μ^+, $A_D(t)$ is the time-dependent asymmetry of the time-delayed diamagnetic μ^+, ν_1 is the spin rotation frequency around H_1, and α is the frequency spread mainly due to the inhomogeneity of the H_1 field (Morozumi et al.,1986).

Further direct information can be obtained by applying a time-delayed r.f. field. By changing the timing of the r.f. pulse with respect to the muon pulse, one can measure the time evolution of the muon state in cases such as Mu $\rightarrow \mu^+$.

6.6.4 Advanced LCR with continuous beam

In the case of a continuous-beam experiment, in order to obtain a LCR signal efficiently, the so-called integral method is helpful. There, without recording a whole μSR time spectrum associated with each incoming muon, the time-integrated asymmetry in B and F counters is recorded as a function of the applied external field:

$$A\bar{G}_z = \frac{N_B - N_F}{N_B + N_F} = \frac{A}{\tau_\mu} \int_0^\infty \exp(-t/\tau_\mu) G_z(t) dt$$

The measurement proceeds without reference to the timing of the incoming muon so that the counting rate can increase in a rate-unlimited way. The LCR signal can be seen as a fraction of decrease in $A\bar{G}_z$.

REFERENCES

Abragam, A. (1984). *C. R. Acad. Sci. Paris*, **Ser. 2**, 95.
Friedman, J.I. and Telegdi, V.L. (1957). *Phys. Rev.*, **106**, 1290.
Garwin, R.L. et al.(1957). *Phys. Rev.*, **105**, 1415.
Hayano, R.S. et al.(1979). *Phys. Rev.*, **20B**, 850.
Heffner, R. H. et al. (2003). *Physica B*, 326.
Heming, M. et al.(1989). *Chem. Phys.*, **129**, 335.
Ivanter, I.G. and Smilga, V.P. (1969). *Sov. Phys. JETP*, **28**, 796.
Ivanter, I.G. et al.(1972). *Sov. Phys. JETP*, **35**, 9.
Jeong, J. et al. (2002). *Phys. Rev.*, **B66**, 13241.
Kadono, R. et al.(1989). *Phys. Rev.*, **B39**, 23.

Karlsson, E.B. (1995). *Solid State Phenomena as seen by Muons, Protons, and Excited Nuclei.* Oxford: Oxford Science Publications.

Kiefl, R.F. and Kreitzman, S.R. (1992). In *Meson Science*, ed. T. Yamazaki, K. Nakai, and K. Nagamine, p. 265. Amsterdam: North Holland.

Kiefl, R.F. *et al.* (1986). *Phys. Rev.,* **A34**, 681.

Kitaoka, Y. *et al.* (1982). *Hyperfine Interactions,* **12**, 51.

Kojima, K. (1995). PhD Thesis. Tokyo: University of Tokyo.

Kreitzman, S.R. *et al.* (1986a). *Phys. Rev. Lett.,* **56**, 181.

Kreitzman, S.R. *et al.* (1986b). *Hyperfine Interactions,* **31**, 13.

Kubo, R. (1981). *Hyperfine Interactions,* **8**, 731.

Kubo, R. and Toyabe, T. (1967). *Magnetic Resonance and Relaxation*, ed. R. Blinc, p. 810. Amsterdam: North Holland.

Minaichev, E.V. *et al.* (1970). *Soviet Physics JETP,* **31**, 849.

Morozumi, Y. *et al.* (1986). *Phys. Lett.,* **A118**, 93.

Nagamine, K. (1981). *Hyperfine Interactions,* **8**, 787.

Nagamine, K. *et al.* (1997). *Hyperfine Interactions,* **104–6**.

Nishiyama, K. (1992). In *Meson Science*, ed. T. Yamazaki, K. Nakai, and K. Nagamine, p. 199. Amsterdam: North Holland.

Roduner, E. *et al.* (2000). *Physica,* **B289–90**.

Schatz, G. and Weidinger, A. (1996). *Nuclear Condensed Matter Physics; Nuclear Methods and Applications.* Chichester: John Wiley.

Schenck, A. (1985). *Muon Spin Rotation Spectroscopy.* Bristol: Adam Hilger.

Shimokoshi, F. *et al.* (1990). *Nucl. Instr.,* **A297**, 103.

Slichter, C.P. (1965). *Principles of Magnetic Resonance.* New York: Harper International.

Sulaiman, S. B. (1994). *Phys. Rev.,* **B48**, 9879.

Uemura, Y.J. (1981). *UT-MSL Report* no. 20.

Yamazaki, T. (1979). *Hyperfine Interactions,* **6**, 115.

7

Muon spin rotation/relaxation/resonance: probing microscopic magnetic properties

7.1 Application of μSR to studies of the intrinsic properties of condensed matter

The applications of muon spin rotation/relaxation/resonance (μSR) to condensed-matter studies can be roughly categorized into two types:

1. The probing of microscopic magnetic properties of the target material which are essentially unchanged by the muon's presence. In this case, the most important features of the experiment are the muon's capability to measure magnetic properties under zero external field, its unique response to spin dynamics, its sensitive detection of weak and/or random microscopic magnetic fields, and so forth. We may call this type of μSR "passive" probe.
2. The creation of a new microscopic condensed-matter system in the target material by the introduction of μ^+, Mu, or μ^-, and the study of the unique microscopic response of the material. Here, representative central topics are the presence or localization of the μ^+ at the interstitial sites and its diffusion properties, electronic properties around μ^+/Mu, the chemical reactions undergone by the hydrogen-like Mu center in semiconductors, transport phenomena of the electron brought in and probed by the μ^+ in conducting polymers and biomolecules, and so forth. We may call this type of μSR "active" probe.

Activity in the field of μSR applications in condensed-matter studies has been increasing since the late 1970s, in parallel with the progress of intense proton accelerator facilities (known as meson factories) such as Los Alamos Meson Physics Facility (LAMPF), Paul Scherrer Institute (PSI), and Tri-University Meson Facility (TRIUMF). The introduction of pulsed muon sources in the early 1980s at High Energy Accelerator Research Organization (KEK) and later at Rutherford Appleton Laboratory (RAL) led to further increases and diversification in μSR activities. Some of the major research highlights covering the period up to the end of the 1990s are summarized in Table 7.1.

In this chapter, we would like to focus on category 1 above, namely studies of the intrinsic properties of condensed matter, under the assumptions that those properties are unchanged by the introduction/presence of the muon itself. Then, in Chapter 8, we will further explore category 2, namely studies of microscopic physics and chemistry created by the introduction of the muon.

Understanding the fundamental properties of matter, in particular, new materials with new functions, makes it possible for us to promote the production of new materials and their

Table 7.1 List of major muon spin rotation/relaxation/resonance (μSR) investigations on condensed matter

1. Measurements of hyperfine structure of the μ^+ at interstitial sites and the bound μ^- just beside the nucleus in magnetic materials
2. Probing magnetic order and spin dynamics such as critical phenomena in magnetic materials like spin glass, heavy fermion systems, low-dimensional systems and exotic magnetic materials
3. Probing high-T_c superconductors and related materials in terms of their magnetic-phase diagram, penetration depth, and flux lattice structure
4. Study of diffusion phenomena of interstitial μ^+ in metals, semiconductors, and insulators
5. Study of structure and reaction of muonium and μ^+ in semiconductors and insulators
6. Study of structure and reaction of muonic radicals and μ^+ involving molecules in chemical systems
7. Study of electron transport in polymers and macromolecules using labeled electrons with μ^+

application to the development of various aspects of human life. Systematic understanding of matter can sometimes only be obtained by learning physical and chemical properties at the microscopic level. As described in Chapter 6, some specific aspects of microscopic properties can be uncovered exclusively by μSR; without μSR, these are masked or unobservable. Therefore, the application of μSR to studies of the intrinsic properties of condensed matter (category 1) is quite important. Since category 1 is the main area of interest in the activities of the entire μSR community, it is impossible to cover all research topics. Therefore, some representative sketches from the present author's related field are presented here. For further details, please refer to proceedings of the regular μSR conferences as mentioned in Chapter 6, as well as some review articles (Schenck and Gygax, 1995; Dalmas de Reotier and Yaouanc, 1997; Kalvius *et al.*, 2001).

7.1.1 Determination of the μ^+ site in solids

The location of the μ^+, which is the charge species most commonly used in "passive" probe studies of host materials, must be determined or at least predicted before the development of any serious arguments based on experimental results.

The nature of the probe should be determined without using the properties to be probed. Therefore the microscopic crystal site of the μ^+ should be found without making use of the host magnetic properties which are the object of the study. The only consistent way of doing this is to use μ^+ spin relaxation due to the surrounding nuclear magnetic moments in the paramagnetic phase where fluctuating electronic moments do not contribute (at least, do not contribute significantly) to μ^+ relaxation. The most relevant example of this type of measurement is the μ^+ in nonmagnetic metals.

Historically, the first example of determination of the μ^+ location was for μ^+ in face-centered cubic (fcc) copper (Camani *et al.*, 1977). In this case, the transverse dipolar broadening in transverse field–muon spin rotation (TF-μSR) was studied as a function of external TF. The influence of the electric field gradient due to the charged muon "impurity" upon the Cu nuclear dipolar coupling (via quadrupolar interactions) was recognized as an important correction and incorporated in the analysis (Hartmann, 1977). In conclusion, the μ^+ was determined to be located at the octahedral site by the analysis of the field-dependent decoupling pattern of the nuclear dipolar interaction perturbed by electric field gradient, allowing for a 5% displacement of the Cu nearest neighbors due to impurity-induced lattice dilation.

In some fortunate situations, the atomic dipolar field can be used to determine the μ^+ location straightforwardly. In the case of ferromagnetic hexagonal close-packed (hcp) Co or Gd, a unique dipolar field is expected at each interstitial site, which is known to be different in sign depending upon whether it is the T-site or the O-site that is occupied, and to change sharply as a function of temperature due to a directional change of the easy axis. Therefore, without entering into the details of electronic structure underlying the atomic moment, the μ^+ location can be determined to be the O-site (Nishida *et al.*, 1978).

Let us next describe how one can learn the basic microscopic magnetic properties of a novel and complex system using the μSR method, taking μ^+SR studies on the high-T_c material $La_{1-x}Sr_xCuO_4$ (LSCO) as an example, the results regarding which are summarized in Figure 7.1 (Torikai *et al.*, 1993).

μ^+ *Location determination*

Using a single crystal of LSCO in its paramagnetic phase, the μ^+ location was determined by measuring the crystal-axis dependence of the nuclear dipolar broadening to be 1 Å (0.1 nm) below the apical oxygen.

Regarding the μ^+ site in the other high-T_c superconductors, the level-crossing resonance (LCR) was applied to $YBa_2Cu_3O_7$ (YBCO) enriched with ^{17}O (Brewer *et al.*, 1990). The result is consistent with the μ^+ being 1 Å (0.1 nm) away from a single oxygen nucleus. The interpretation of the data is still uncertain, including the possibility of multiple μ^+ sites.

Again as a general example the anisotropic Knight shift for μ^+ in paramagnetic materials can be used towards the μ^+ location determination if a classical dipolar contribution from the surrounding moments is assumed (e.g., Schenck *et al.*, 2002).

μ^+ *Hyperfine field and nature of the electron spin system*

Taking a well-understood system as an analog, the local field vector (strength and direction) at the μ^+ should be studied to learn the nature of the hyperfine interactions between the localized μ^+ and the surrounding electronic spin density distribution. In this case, the μ^+ field vector and the spin of the Cu moment were determined from measurements on La_2CuO_4. These data, together with the knowledge of the μ^+ location, can be used as a basis for further considerations.

Studies of the magnetic-phase diagram

The hole- or electron-doping dependence can be studied using a series of samples with varying dopant concentrations; since the systems are structurally analogous, the μ^+ location might be assumed to remain almost unchanged, while the μ^+ hyperfine field changes systematically with respect to the undoped system. The regions of the concentration-phase diagram where static or dynamic magnetic order occurs can be studied. The magnetic properties of LSCO have been studied in this way as a function of hole concentration (x).

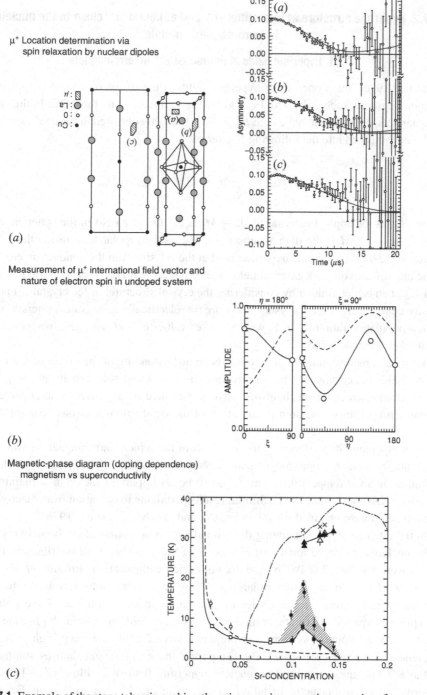

Figure 7.1 Example of the steps taken in probing the microscopic magnetic properties of new materials by the μ^+SR, taking the case of single crystalline LSCO as an example: (a) determination of the μ^+ site; (b) studies of hyperfine field vector of the well-known undoped La_2CuO_4; and (c) exploration of the new magnetic-phase diagram by variation of the Sr dopant concentration.

7.2 Hyperfine structure at interstitial μ^+ and at bound μ^- close to the nucleus in ferromagnetic metals

7.2.1 Hyperfine fields at interstitial μ^+ in ferromagnets

In the late 1970s, the properties of hyperfine fields at the μ^+ were studied in typical ferromagnetic metals such as Ni, Co, Fe, and Gd, as summarized in Table 7.2. In the case of ferromagnets, the local magnetic field B_μ at the μ^+ directly obtained in the μSR experiment can be decomposed into the following five terms:

$$B_\mu = H_{ext} + \frac{4\pi}{3} M_s + DM + H_{dip} + H_{int}$$

where H_{ext} is the applied external field, $\frac{4\pi}{3} M_s$ is the Lorentz field in the spherical cavity around the μ^+, DM is the diamagnetic field, H_{dip} is the dipolar field inside the Lorentz sphere, and H_{int} is the contact hyperfine field at the μ^+ site from the conduction electrons. In the case of zero or weak external field, one considers that $H_{ext} + DM \cong 0$. The dipolar field H_{dip} can be determined by considering the crystal structure: in fcc crystals, it is zero; in body-centered cubic (bcc) crystals, there are two electrically inequivalent interstitial sites (with a population ratio of 2 to 1) where it takes values of $-H_1$ and $2H_1$, respectively; in hcp it takes a unique value.

These data, representing research towards an understanding of the origin of the μ^+ hyperfine fields, taken together with corrections for enhanced spin density at the μ^+ site due to charge-screening mechanisms, have contributed to a clearer general picture of the basic properties of the spin polarization of the conduction electrons near interstitial sites.

The temperature dependence of the μ^+ field in the typical ferromagnet Ni was found to deviate from macroscopic magnetization (Nagamine et al., 1976). Stoner-type electron excitation at finite temperatures was found to be an important factor in explaining the deviation (Kanamori et al., 1981). Similar results pertaining to deviation from macroscopic magnetization were obtained for μ^+ in Fe, Co, and Gd (Nishida et al., 1978).

On the other hand, μ^- occupying the ground state of a muonic atom, bound very close to the nucleus, can probe local spatial aspects of the hyperfine field distribution. Experiments were conducted at PSI by the Tokyo–Zurich collaboration (Imazato et al., 1984; Keller et al., 1986) in which polarized μ^- were injected into ferromagnetic samples under zero external field. Under these circumstances, two-thirds of the polarized μ^- spin is expected to rotate around the hyperfine field. A special high time resolution decay-e$^-$ counter system was prepared to deal with the very high precession frequencies expected (> 1 GHz). By comparing the experimental data as summarized in Table 7.2 to the corresponding nuclear hyperfine field of a dilute $(Z - 1)$ nucleus impurity in a ferromagnetic metal of atomic number Z (such as Mn in Fe, Co in Ni, etc.), detailed information regarding the electronic core polarization implicated in the origin of the nuclear hyperfine fields in these systems was obtained (Yamazaki, 1981).

Table 7.2 Summary of μ^+ and μ^- hyperfine fields in typical ferromagnets

	Probable location	B_μ (Gauss)	B_{dip} (Gauss)	H_{int} (Gauss)	M_s (Gauss)	$H_{int}/\frac{8\pi}{3}M_s$	M_{int}/M_s (neutron data)	$H_{int}/\frac{8\pi}{3}M_{int}$
Fe (bcc)	Tetrahedral	−3773(10) 73.4 K	~0	−11100	1749	−0.76	−0.074	10
Co (hcp)	Octahedral	−318.5(4) 4.2 K	127	−6100	1458	−0.51	−0.61	3.0
Ni (fcc)	Octahedral	+1495.6(21) 0.1–4.2 K	0	−641	510	−0.15	−0.15	1.0
Gd (hcp)	Octahedral	+1083(3) 4.2 K	373	−7400	2115	−0.42	−0.12	3.4

bcc, body-centered cubic; hcp, hexagonal close-packed; fcc, face-centered cubic.

	B_μ (T)	$B_{Lorentz}$ (T)	B_μ^{hf} (T)	$B_{N(Z-1)}^{hf}$ (T)	Δ^a (%)
μ^- Ni	−11.63 (at 23 K)	0.21	−11.84	−12.13[b]	−2.4(3)
μ^- Fe	−18.3 (at 323 K)			−18.41[c]	−0.6(3)

[a] $(B_\mu^{hf} - B_N^{hf})/B_N^{hf}$.
[b] ^{59}Co in Ni nuclear magnetic resonance data.
[c] ^{55}Mn in Fe nuclear magnetic resonance data.

7.3 Probing critical phenomena and magnetic ordering in metal ferromagnets and heavy fermions

The μSR technique can be applied to probe static or dynamic magnetic ordering in magnetic materials under zero external field and at high temperatures; this is useful for the understanding of critical phenomena just above or below the critical temperature. In this respect μSR is much more convenient than nuclear magnetic resonance (NMR) which, in most cases, needs an external field and cannot easily be applied at high temperatures.

Critical phenomena have been studied extensively for μ^+ in ferromagnetic Ni (Nishiyama et al., 1984) and Gd (Wäckelgård et al., 1986); in each case a critical index typical for three-dimensional Heisenberg ferromagnets was obtained.

Regarding itinerant ferromagnets such as MnSi, the characteristic temperature-dependent relaxation rates $\lambda \propto T/(T - T_c)$ predicted by the theory of Moriya (Moriya and Ueda, 1974) were confirmed for the first time by μSR (Hayano et al., 1978). The muon result stands as a complement to the NMR work which, because of limitations on the sensitivity time range, observed only the high-temperature region ($T >> T_c$) of the range predicted by theory.

Some compounds containing rare-earth elements such as Ce or Yb or actinide elements such as U are known to exhibit charge carriers with effective masses substantially heavier than a bare electron. Such heavy fermion systems originate from the interaction between localized f-electron spins and conduction-electron spins. At low temperature, some of these species exhibit magnetic ordering and/or superconductivity. The sensitive nature of μSR has been made use of to probe the ordering of small magnetic moments, for example those (typically less than 0.1 μ_B) appearing in typical heavy fermion systems such as $CeAl_3$ or UPt_3, as summarized in a review paper (Amato, 1997). However, because of the strong sensitivity of the μSR to magnetic impurities, some unknown factors exist regarding the magnetic ordering; improved samples may produce different μSR results. In the case of UPt_3, contradictory experimental results exist regarding magnetic order in superconducting state between neutron scattering and μSR; magnetic ordering is seen by neutrons at $T_N \cong 6$ K, while no order is seen by μSR, suggesting a fluctuating magnetic order reflecting a difference in the time window for the fluctuation time constant between neutron scattering and μSR (Higemoto et al., 2000). The neutron scattering does see a rapidly fluctuating moment in the time range around 10^{-12} s, while the μSR sees a fluctuating moment in the time range from 10^{-6} s to 10^{-9} s; the magnetic order appears statically in the time range shorter than 10^{-9}. Careful measurements of μ^+ Knight shift showed two distinct isotropic signals for the field in basal plane around T_N, implying two component magnetic response in UPt_3 (Yaouanc et al., 2000).

7.4 Probing spin dynamics in random and/or frustrated spin systems

The μSR has a unique response to the dynamic behavior of the surrounding magnetic moments, as seen in Figure 6.6. Also, it is not necessary to have a coherent nature in spin dynamics in order to obtain a μSR signal. Therefore, spin dynamics in random magnets such as spin glasses have been amenable to study using the μSR technique.

Systematic μ^+SR studies on the universal temperature dependence of the correlation time have been pursued on spin glass systems of noble metals with dilute magnetic impurities, such as CuMn or AuFe (Uemura et al., 1985). This type of μSR study has been extended to probe spin dynamics in mixed-ordered magnets which exhibit competing magnetisms such as competing anisotropy like $Fe_{1-x}Co_xTiO_3$ (Torikai et al., 1994), or in which ferromagnetism and antiferromagnetism compete, as in $Fe_{1-x}Mn_xTiO_3$. In these systems, the spin dynamics remaining in the unfrozen component were clearly observed by μSR, thanks to the unique range in which the correlation time lies; these measurements would not have been easily accomplished by either Mössbauer spectroscopy or neutrons.

The typical example of frustrated spin systems is a triangular lattice. The magnetic ion occupying one of the three sites is subject to frustration; after two spins taking ferro- or antiferro-coupling, the third spin always has frustration concerning its spin direction. Various experimental studies have been undertaken using μSR to explore spin dynamics of these frustrated spin systems, e.g. so-called Kagomé lattice system, Pyrochrore antiferromagnet, by taking advantage of the method's unique time window for the spin dynamics (Uemura et al., 1994, Gardner et al., 1999, Keren et al., 2000).

7.5 Probing magnetism, penetration depth, and vortex states in high-T_c superconductors

After the discovery of high-T_c superconductors in 1986, because of their potential applications, the μSR activities were devoted to various types of experimental investigations on the microscopic understandings of the origin of superconductivity.

7.5.1 Magnetism in high-T_c superconductors

Making use of the advantages of μSR – which is a microscopic magnetic probe which can be used under zero external field and that it is sensitive to weak and/or random magnetic order – the method has been profitably applied to probe magnetism in high-T_c superconductors since the discovery of these materials in 1987. From the earliest days of the high-T_c superconductor, the interplay between superconductivity and magnetism has been known to be a central feature of the basic physics of CuO_2 materials.

Historically, the usefulness of μ^+SR methods for studying the magnetic properties of high-T_c superconductors initially came to light as a result of the discovery of antiferromagnetic ordering in the oxygen-reduced Tetra II phase of YBCO (Nishida et al., 1987). In that system, as seen in Figure 7.2, the μ^+SR spectrum was measured for the YBCO crystal as a function of the oxygen concentration (x): no coherent μ^+ spin precession was detected and only μ^+ relaxation due to nuclear moments was observed in both the Ortho-I ($x > 6.8$) and Ortho-II ($6.4 < x < 6.8$) regions; however, a coherent

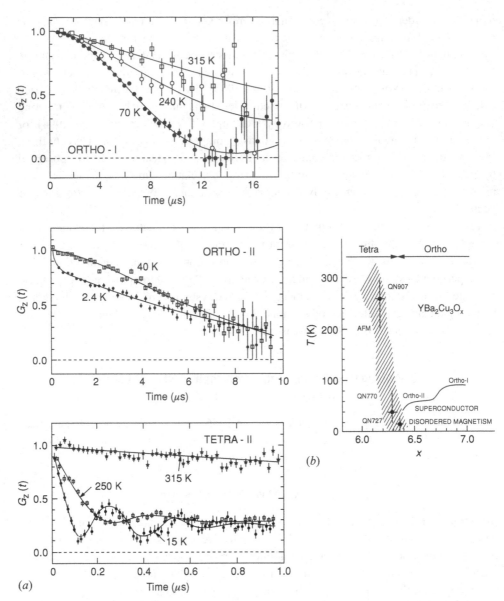

Figure 7.2 The μ^+SR time spectrum observed in the first trials on magnetic ordering studies of the high-T_c superconductor $YBa_2Cu_3O_7$ (YBCO) for the three crystal structures corresponding to different oxygen concentration regions (Nishida *et al.*, 1987) (*a*), and the magnetic phase diagram obtained (*b*).

μ^+ precession corresponding to an internal field of 300 G was detected for the Tetra-II ($x < 6.2$) region.

This μSR experiment on YBCO aroused considerable interest in the application of the μSR method to magnetic-phase measurements on high-T_c-related materials. Salient examples of this work are YBCO, discussed above, BSCCO ($Bi_2Sr_2(Ca_{1-x}Y_x)Cu_2O_{8+\delta}$) (Nishida et al., 1988), and LSCO, which will be mentioned further below.

The second example to be given here is related to the so-called "stripe order or 1/8 problem" in high-T_c superconductors, in which experimental progress has followed the step-by-step development below.

In the $La_{2-x}Ba_xCuO_4$ cuprate system, for the $0.05 \leq x \leq 0.28$ region in which the system is a superconductor, T_c suppression was found to occur in a narrow region around $x = 0.125$ (1/8), where there is a crystal structure change from low-temperature orthorhombic (LTO) to low-temperature tetragonal (LTT) at temperatures above T_c. Antiferromagnetic order was found to exist in this $x = 0.125$ region first by μSR (Kumagai et al., 1991), then later by NMR and neutron scattering.

In the LaSr system, if 40% of the La is replaced by Nd, at $x = 0.125$ and below the LTO to LTT structure change which accompanies the suppression of T_c, neutron scattering has observed antiferromagnetic ordering and the existence of a stripe structure of charge and spin (Tranquada et al., 1995).

In the pure LSCO system (i.e., without any replacement of Cu by other elements), which does not undergo any crystal structure change, a spin-glass order with a slight T_c suppression has been suggested to occur at $x = 0.115$ (Torikai et al., 1990, 1992), while antiferromagnetic ordering with T_c suppression has been observed at $x = 0.115$ in another measurement (Watanabe et al., 1994). Also, in recent neutron scattering, NMR, and supersonic measurements at $x = 0.12$, the existence of a magnetic superlattice, magnetic order in the La nuclei, and a softening of the longitudinal sonic wave have been detected, suggesting a dynamic pinning of spin correlations by a lattice instability towards LTT (Suzuki et al., 1998).

Recently the anomaly in LSCO at $x = 0.115$ has been revisited. By comparing the two existing sets of data on the two original samples to the data obtained on a new refined sample made by the Tokyo University of Science group (Arai et al., 1999) we obtained Figure 7.3. Here we summarize the observations: (1) there is antiferromagnetic ordering at $x - 0.115$ with a transition temperature of around 10 K, no matter whether the system is superconductive or not; (2) the occurrence of antiferromagnetic ordering is common to all three samples.

All of these results suggest that, at $x = 0.115$, instead of $x = 0.125$, the LSCO system may undergo a unique antiferromagnetic ordering. By introducing spin defect by replacing a small amount ($\sim$1%) of Cu with Zn, it is well known that a static stripe order becomes stabilized. There, for example, the antiferromagnetic (AF) phase can be seen in the case of $x = 0.125$.

The 1/8 problem, in addition to the La-based high-T_c oxides, seems also to exist in the BSCCO system, where dynamic relaxation of the μ^+ spin was seen at 0.3 K for Ca- and Zn-doped samples with effective hole doping of $x = 0.125$, as shown in Figure 7.4 (Koike et al., 1999; Watanabe et al., 1999, 2000). A similar phenomenon was seen in $YBa_2Cu_{3-2y}Zn_{2y}O_{7-\delta}$ at around a hole concentration of 1/8 per Cu (Akoshima et al., 2000). The present state of knowledge related to the 1/8 problem is summarized in Table 7.3.

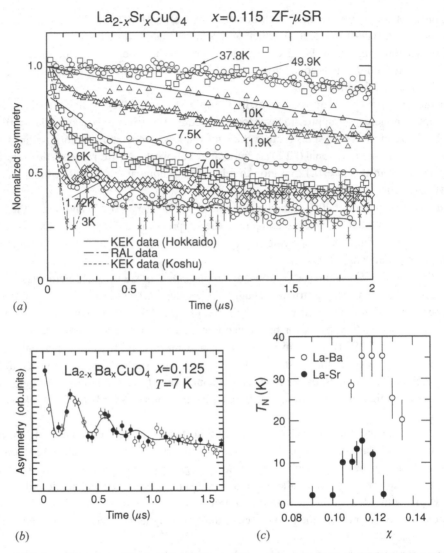

Figure 7.3 The μSR patterns obtained for the three different $La_{2-x}Sr_xCuO_4$ (LSCO) samples with $x = 0.115$ under zero external field where KEK data (Koshu) is taken from Torikai *et al.*, 1992, KEK data (Hokkaido) is taken from Watanabe *et al.*, 1994 and RAL data is taken from Arai *et al.*, 1999. (*a*). The corresponding data for $La_{2-x}Ba_xCuO_4$ (LBaCO) with $x = 0.125$ (Kumagai *et al.*, 1991) (*b*). The transition temperature T_N versus hole concentration in both LBaCO and LSCO cases (Watanabe *et al.*, 1994) (*c*). KEK, High Energy Accelerator Research Organization; RAL, Rutherford Appleton Laboratory; ZF, zero field.

Throughout the μSR experiments on almost all the high-T_c superconductors studied, it is now established that the 1/8 anomaly exhibiting a stripe order is a general property of the CuO_2 plane in all the high-T_c oxide superconductors. The result can be extended to cover all the hole concentrations by introducing a time domain of the dynamical stripe correlation.

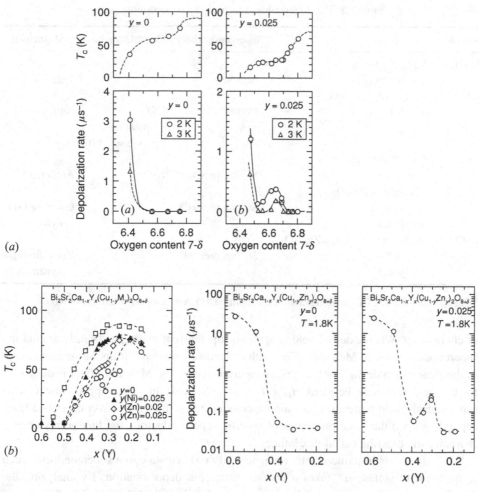

Figure 7.4 Examples exhibiting the stripe magnetic order as well as suppression of superconductivity in representative high-T_c superconductors, all of which are demonstrating an enhanced appearance of magnetic order with Zn substitution for Cu. (a) $YBa_2Cu_{3-2y}Zn_{2y}O_{7-\delta}$ with $y = 0$ and 0.025 (Akoshima *et al.*, 2000); and (b) $Bi_2Sr_2Ca_{1-x}Y_x(Cu_{1-y}M_y)_2O_{8+\delta}$ with $y = 0$ by replacing 0.0, 0.01 (Watanabe *et al.*, 2000) and $y(Zn) = 0.025$ (Koike *et al.*, 1999).

There is a theory suggesting that a dynamical stripe correlation is the possible mechanism producing high-T_c superconductivity (Kivelson *et al.*, 1998).

7.5.2 Penetration depth and vortex states in high-T_c superconductors

When external magnetic field (H) and electric current (J) are applied to a superconductor, superconductivity exists in a region of (T, H, J) below critical values of (T_c, H_c, J_c). In this superconductivity region, there are several characteristic phenomena: (1) Meissner effect ($B = 0$); (2) existence of persistent current; (3) Josephson effect; and (4) existence

Table 7.3 The 1/8 problem in high-T_c superconductors

Sample		Superconductivity	Crystal structure	Magnetism
LaBa	$La_{2-x}Ba_xCuO_4$			
	$x = 0.125(1/8)$	Non-Sc	LTT	Antiferro
La(Nd)Sr	$La_{1.6-x}Nd_{0.4}Sr_xCuO_4$			
	$x = 0.125(1/8)$	Non-SC	LTT	Antiferro
			Spin/charge	
			Stripe structure	
LaSr	$La_{2-x}Sr_xCuO_4$			
	$x = 0.115$	SC (suppressed)	LTO	Antiferro
YBCO	$YBa_2Cu_{3-2y}Zn_{2y}O_{7-\delta}$			
	$c_{hole} = 0.125$	SC (suppressed)		Correlated spin
	$y = 0.025$			dynamics
BSCCO	$Bi_2Sr_2Ca_{1-x}Y_x(Cu_{1-y}Zn_y)_2O_{8+\delta}$			
	$c_{hole} = 0.125$	SC (suppressed)		Correlated spin
	$y = 0.025$			dynamics

SC, superconductor; LTT, low-temperature tetragonal; LTO, low-temperature orthorhombic.

of energy gap. When external field is applied, supercurrent exists persistently around the superconductor due to Meissner effect with a characteristic depth from the surface known as the penetration depth. For the type II superconductor, the Meissner effect is destroyed for the field region H between H_{c1} (lower critical field) and H_{c2} (upper critical field) and magnetic field penetrates into the superconductor as flux vortex. As described later, depending upon the basic properties of specific superconductors, the flux vortex takes a characteristic ordering called flux lattice.

When the μ^+ is introduced into a magnetically ordered state in a superconductor such as a magnetic vortex, μ^+ takes a characteristic spin depolarization. By analyzing the depolarization pattern, it is possible to explore the details of the behavior of vortices. Since the magnetic length of the vortex is larger than the atomic scale, the μ^+ can randomly probe or sample the behavior of the vortex.

The penetration depth λ is related to the average width of inhomogeneous field ΔB by $\Delta B \propto \lambda^{-2}$. The field inhomogeneity ΔB affects μ^+ depolarization appearing in $G_x(t) = \exp(-\sigma^2 t^2/2)$ where $\sigma \propto B$. The penetration depth λ, in a clean-limit superconductor where ℓ (mean free path) $>> \xi$ (coherence length), is related to superconducting carrier density n_0 and effective mass m^* as:

$$\sigma \propto \frac{1}{\lambda^2} = \frac{4\pi n_s e^2}{m^* c^2}$$

Since discovery of a relationship between σ (relaxation rate in µSR time spectrum and n_s/m^*, the data compilations on T_c versus $\sigma(T \to 0)$ throughout almost all the type II superconductors have been done by Uemura et al. (1991, 1997), and universal correlations

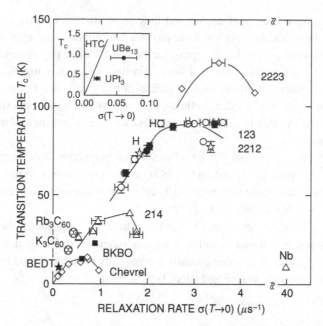

Figure 7.5 The universal correlation between T_c and the transverse field–muon spin rotation (TF-μSR) depolarization σ in various superconducting materials (Uemura et al., 1991).

between T_c and n_s/m^* were found (Figure 7.5). There, several important observations have been made: (1) T_c is determined by the value of n_s/m^*; (2) by converting n_s/m^* to effective Fermi temperature T_F, which represents the energy scale for superconducting carrier translational motion, the $T_c - T_F$ relationship can classify the high-T_c superconductor cuprates, including some exotic superconductors, against conventional Bardeen–Cooper–Schrieffer (BCS) superconductors ($T_F/T_c << 0.01$ in the latter case).

A type II superconductor, under a sufficiently large applied field, shows spatial regions of field inhomogeneity called vortices. When the many vortices exist inside the type II superconductors, the electromagnetic interaction between vortices arrange themselves to take a two-dimensional periodic lattice. Such a vortex lattice produces a characteristic field distribution $P(B)$ of the internal flux density, which can cause a characteristic damping of the μSR time spectrum under an applied transverse field. The probability distributions of this type have been observed by various μSR experiments (Herlach et al., 1990; Lee et al., 1993), followed by other experiments observing "melting of vortex lattice," details of lattice structure with reference to vortex core structure, etc. (Sonier et al., 2000).

7.6 Probing magnetic ordering in exotic magnetic materials

In order to understand the origin of superconductivity in high-T_c superconductors, various new types of magnetic materials which have in common the nature of strongly correlated electron systems have been synthesized and they were studied by various experimental methods, including μSR. One typical example is a low-dimensional spin systems of the types variously known as spin-ladder, spin-Peierls, and Haldane gap systems. Contrary to the classical view, such a system does not show spin-freezing or ordering. Several magnetic systems with a spin-singlet ground state which is nonmagnetic due to the existence of a spin gap (the gap is known by other methods, e.g., inelastic neutron scattering) have been extensively studied by μSR because of its sensitivity to the occurrence of magnetic order in the nonmagnetic state.

The μ^+SR studies on spin-singlet one-dimensional magnetism have produced the following results. The spin-Peierls system $CuGeO_3$, as expected, shows no intrinsic magnetic ordering, while magnetism induced by impurity effects in this system has been demonstrated and explored by μ^+SR (Kojima et al., 1997).

The two-leg spin-ladder compound $LaCoO_{2.5}$, which was expected to be nonmagnetic at low temperatures, was found to undergo antiferromagnetic ordering at 125 K (Kadono et al., 1996); this phenomenon was eventually theoretically explained by taking into account interladder interactions (Normand and Rice, 1997).

REFERENCES

Akoshima, M. et al. (2000). Phys. Rev., **B62**, 6761.

Amato, A. (1997). Rev. Mod. Phys., **69**, 1119.

Arai, J. et al. (1999). J. Low Temp. Phys., **117**, 377.

Brewer, J.H. et al. (1990). Hyperfine Interactions, **63**, 177.

Camani, M. et al. (1977). Phys. Rev. Lett., **39**, 836.

Dalmas de Reotier, P. and Yaouanc, A. (1997). J. Physics/Condens. Matter, **9**, 9113.

Gardner, J.S. et al. (1999). Phys. Rev. Lett., **82**, 1012.

Hartmann, O. (1977). Phys. Rev. Lett., **39**, 832.

Hayano, R.S. et al. (1978). Phys. Rev. Lett., **41**, 1743.

Herlach, D. et al. (1990). Hyperfine Interactions, **63**, 41.

Higemoto, H. et al. (2000). Physica B, **281 & 282**, 984.

Imazato, J. et al. (1984). Phys, Rev. Lett., **53**, 1849.

Kadono, R. et al. (1996). Phys. Rev., **B54**, R9628.

Kalvius, G.M., Noakes, D.R., and Hartmann, O. (2001). Handbook on the Physics and Chemistry of Rare Earths, ed. K.A. Gschneidner, Jr., L. Eyring, and G.H. Lander, p. 55. Amsterdam: North Holland.

Kanamori, J. et al. (1981). Hyperfine Interactions, **8**, 573.

Keller, H. et al. (1986). Hyperfine Interactions, **31**, 461.

Keren, A. et al. (2000). Phys. Rev. Lett., **84**, 3450.

Kivelson, S.A. et al. (1998). Nature, **393**, 550.

Koike, Y. et al. (1999). Int. J. Mod. Phys., **B13**, 3546.

Kojima, M.K. et al. (1997). Phys. Rev. Lett., **79**, 503.

Kumagai, K. et al. (1991). Physica C, **185–9**, 913.

Lee, S.L. et al. (1993). Phys. Rev. Lett., **71**, 3262.

Moriya, T. and Ueda, K. (1974). Solid State Comm., **15**, 169.

Nagamine, K. et al. (1976). Hyperfine Interactions, **1**, 517.

Nishida, N. *et al.* (1978). *J. Phys. Soc. Jpn*, **44**, 1131.

Nishida, N. *et al.* (1987). *Jpn J. Appl. Phys.*, **26**, L1856.

Nishida, N. *et al.* (1988). *Physica*, **C156**, 625.

Nishiyama, K. *et al.* (1984). *Hyperfine Interactions*, **17–19**, 473.

Normand, B. and Rice, T.M. (1997). *Phys Rev.*, **B56**, 8760.

Schenck, A. and Gygax, F.N. (1995). *Handbook of Magnetic Materials*, vol. 9, ed. K.H.J. Bushow, p. 57. Amsterdam: North Holland.

Schenck, A. *et al.* (2002). *Phys. Rev.*, **B66**, 144404.

Sonier, J.E. *et al.* (2000). *Rev. Mod. Phys.*, **72**, 769.

Suzuki, T. *et al.* (1998). *Phys. Rev.*, **B57**, R3229.

Torikai, E. *et al.* (1986). *Solid State Comm.*, **58**, 839.

Torikai, E. *et al.* (1990). *Hyperfine Interactions*, **63**, 271.

Torikai, E. *et al.* (1992). *Hyperfine Interactions*, **79**, 905, 915, 921.

Torikai, E. *et al.* (1994). *Hyperfine Interactions*, **84**, 105.

Tranquada, J.H. *et al.* (1995). *Nature*, **375**, 561.

Uemura, Y.J. *et al.* (1985). *Phys. Rev.*, **B31**, 546.

Uemura, Y.J. *et al.* (1991). *Phys. Rev. Lett.*, **66**, 2665.

Uemura, Y.J. *et al.* (1994). *Phys. Rev. Lett.*, **73**, 3306.

Uemura, Y.J. *et al.* (1997). *Hyperfine Interactions*, **103**, 35.

Wäckelgård, E. *et al.* (1986). *Hyperfine Interactions*, **31**, 325.

Watanabe, I. *et al.* (1994). *Hyperfine Interactions*, **86**, 603.

Watanabe, I. *et al.* (1999). *Phys. Rev.*, **B60**, R9955.

Watanabe, I. *et al.* (2000). *Phys. Rev.*, **B62**, 14524.

Yamazaki, T. (1981). *Hyperfine Interactions*, **8**, 463.

Yaouanc, A. *et al.* (2000). *Phys. Rev. Lett.*, **84**, 2702.

8

Muon spin rotation/relaxation/resonance: probing induced microscopic systems in condensed matter

As mentioned at the beginning of Chapter 7, there have been two different attitudes in the experimenter's mind concerning the role of the μ^+ in condensed matter, as shown schematically in Figure 8.1(a): at one extreme, the μ^+ is treated as a gentle and passive probe to probe the condensed matter with minimal perturbation and to observe its intrinsic properties prior to the introduction of μ^+; at the other extreme, the μ^+ is treated as a violent and active probe introducing a perturbation in the host material so as to study new physics and chemistry created by the presence of the μ^+. In this chapter, representative studies utilizing the second category, including similar studies of the μ^-, are described.

The role of this type of muon spin rotation/relaxation/resonance (μSR) studies is quite significant in its contribution to the growth of our human daily life: (1) the localization and diffusion of the light hydrogen isotope Mu ($\mu^+ e^-$) simulates a behavior of a dilute hydrogen atom in metals and other condensed matter which is quite difficult to monitor and important in various aspects of industrial constructions; (2) a trace impurity hydrogen-like atom can "passivate" electrical activity of donors and acceptors or "hydrogenate" dangling bonds in semiconductors; (3) the lightest hydrogen atom can explore the most fundamental mechanism of hydrogen chemical reaction in terms of mass dependence; (4) the electron brought in by the energetic light hydrogen can be used to probe electron transport in conducting polymers and biological macromolecules.

8.1 μ^+ localization and diffusion in condensed matter

The μ^+, when it is injected into solid materials like metals, is preferentially located at the interstitial sites, where it causes a dilation of the surrounding lattice atoms due to the effect of screening electrons around the positively charged μ^+ and creates a deep potential well, as shown schematically in Figure 8.1(b). This idea, known as "self-trapping," is a fundamental mechanism essential to the understanding of μ^+ diffusion properties in metals. The system consisting of the μ^+ particle together with the surrounding distorted lattice is referred to as a "small polaron." The theoretical physics of small polarons was developed in the 1950s in relation to the mechanism of electronic conductivity in insulators. Thus, the μ^+ diffusion process in metals can be modeled as the interaction between a small polaron and lattice phonons. At relatively high temperatures μ^+ diffusion can be described in terms of underbarrier hopping with the creation and annihilation of phonons. A series of theoretical

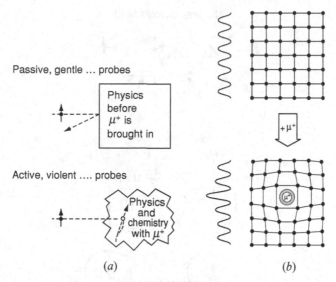

Passive, gentle ... probes

Active, violent .... probes

(a)   (b)

Figure 8.1 (a) Active and passive probe pictures of the μ^+SR method and (b) concept of μ^+ trapping as a self-trapped polaron.

works has been published concerning the transport phenomena of light interstitials in metals based upon the small polaron picture. In metals, it is important to consider an effect of the interaction between the μ^+ and the conduction electrons which screen the μ^+. The behavior of the screening electrons is important for understanding of μ^+ diffusion phenomena at low temperatures. The importance of the electron–muon interaction has been emphasized by Kondo (1984), Yamada (1984), and Kagan (1992). Thus, μ^+ interactions with both phonons and electrons must be incorporated into any theoretical treatments of μ^+ diffusion.

Since the first pioneering experiment carried out by Gurevich *et al.* (1972), diffusion phenomena relating to μ^+ in various metallic materials have been a central topic in the field of experimental μSR. The close relationship with hydrogen in metals has been emphasized, and there have been many discussions on the comparison between the two species.

Now, let us summarize the most important features in the diffusion properties of interstitial μ^+ in metals. All of these characteristics highlight the advantages of muon studies compared with direct experimental research on hydrogen, and illustrate their great usefulness in understanding the diffusion of hydrogen in metals.

1. Isotope effect: the mass difference between μ^+ and H^+ illuminates the greater importance of quantum diffusion phenomena for light particles at low temperatures, while the difference between μ^+ and e^+ illustrates the interplay between band-like diffusion and hopping diffusion.
2. Dilute limit: due to the extremely low concentration of implanted muons, there is no particle–particle correlation at all in the case of the μ^+. Thus, μ^+ diffusion phenomena are free from the "pair" formation effect which is sometimes considered in understanding data on hydrogen diffusion.

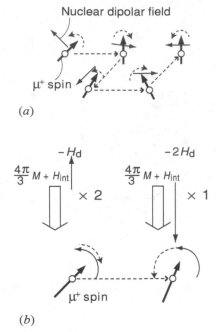

Figure 8.2 Schematic picture of narrowing of the dipolar width due to two typical types of random field sources: (*a*) nuclear dipolar fields and (*b*) atomic dipolar fields at interstitial sites in a ferromagnetic body-centered cubic crystal.

3. No phase problem: μ^+ can be injected into any material producing a homogeneous "α"-phase (the binary system formed between the μ^+ and the host metal), while in the case of hydrogen in metals, phase problems always set limiting conditions on the range of physical parameters which can be studied, and the creation of a pure α-phase at low temperatures is always extremely difficult.

The basic principle of μ^+ diffusion studies is to observe motional narrowing of the dipolar width due to the random magnetic moments appearing in either the transverse or zero/longitudinal relaxation functions. Thus far, the following types of dipolar broadening have been observed in μ^+ diffusion studies (Figure 8.2): (1) nuclear dipolar broadening in various metallic crystals containing nuclei with magnetic moments; (2) atomic electron dipolar broadening, mostly in magnetic materials where the atomic magnetic moments produce randomly oriented static dipolar fields at the interstitial μ^+; the typical example here would be μ^+ in ferromagnetic body-centered cubic (bcc) Fe, where the dipolar fields have different values (H_1, $-2H_1$) at electrically inequivalent interstitial sites; (3) broadening due to magnetic field vortex penetration in type II superconductors, e.g., Nb. The first two of these sources of broadening are distinguished in that, for static dipolar broadening and nondiffusing μ^+, the corresponding relaxation rates take completely different values because of the large difference in the sizes of the moments; these are on the order of a few G for the nuclear dipole, but on the order of kG for the atomic dipole. Thus, the diffusion rates corresponding to a given degree of narrowing are entirely different. Suppose that the observed μ^+SR relaxation rate is between 10^4 and 10^8 s^{-1}; then the jumping frequency ν is from 10^4 to 10^7 s^{-1} in the nuclear case and from 10^8 to 10^{12} s^{-1} in the atomic case.

In the following, we will review the present status of experimental and theoretical studies on the diffusion properties of the μ^+ in metals. We will concentrate our discussions mainly on μ^+ diffusion in pure metals and Mu diffusion in pure ionic crystals, where one can expect the influence of impurities and defects to be at a minimum, and where μ^+ trapping phenomena at impurities can be viewed simply as a means of understanding the intrinsic properties. Some recent review articles exist (Kadono, 1992; Storchak and Prokof'ev, 1998).

Before going into the details of μ^+ diffusion in metals, let us recall briefly the μ^+ history prior to thermalization. There has been some discussion on the possibility that μ^+ may be trapped before it is thermalized. The model is as follows: there might be a metastable μ^+ state extended over several crystal sites; during the deceleration process, the μ^+ might be trapped in this metastable state. Although the idea of metastable state formation is important, there has been no experimental evidence to support this idea. At the same time, there are no theoretical predictions for observable quantities such as dipolar width or hopping rates which can distinguish metastable state formation from muon location at interstitial sites. Therefore, we will not spend any further time on this possibility.

Also, as described in section 3.2, there is no evidence for the formation of any muonium-like paramagnetic state in metals. Therefore, the μ^+ can be treated as occupying a diamagnetic state screened by the incoherent conduction electrons. All the magnetic interactions due to the surroundings come from the bulk magnetic properties of the host metal, somewhat modified by the presence of the μ^+.

8.1.1 μ^+ diffusion in Cu (fcc) and other pure metals

After the first experiment by Gurevich et al. (1972), there were several further measurements of the μ^+ relaxation associated with diffusion properties using the transverse-field method. Through these experiments, impurity effects and lattice deformation effects were studied at temperatures above 4 K, and the diffusion mechanism in this temperature region was revealed to be underbarrier hopping. Using the incoherent tunneling–hopping model of Flynn and Stoneham (1972), the tunneling matrix element was derived to be 18 µeV, with an activation energy of 75 meV, far smaller than that for H in Cu, supporting the underbarrier hopping picture.

Remarkable progress on this system was subsequently made in experiments done at the European Organization for Nuclear Research (CERN), in which the width of the Gaussian damping in the transverse relaxation was seen to have a strange temperature dependence below 20 K (Hartmann et al., 1980); the width has a minimum at around 20 K and increases again with decreasing temperature. In practice, in the transverse relaxation method, it is extremely difficult to determine the dipolar width and the hopping rate separately, while these two quantities are easy to discriminate via longitudinal relaxation under zero external field (ZF). This advantage was then applied to the low-temperature μ^+ diffusion problem in

Cu by Clawson *et al.* (1983). In a ZF experiment, they confirmed that, at low temperatures, the dipolar width is almost constant, while the hopping rate does in fact increase with decreasing temperature below 20 K.

A complete ZF measurement over a wide temperature range was carried out at High Energy Accelerator Research Organization–Meson Science Laboratory (KEK-MSL) employing pulsed μSR (Kadono *et al.*, 1985, 1989). The capability of the pulsed μSR method to measure out to long times is an advantage in studying this type of slow diffusion phenomenon, since the significant pattern change in the spectra appears in the late part of the relaxation function (the 1/3 recovery component described in Chapter 6). A typical ZF μ^+-relaxation spectrum and the extracted dipolar width and hopping rates are presented in Figure 8.3. Several distinctive features were observed in this experiment: (1) the dipolar width is constant within 3% around 0.38 μs^{-1} in the entire temperature range from 20 mK to 200 K, the value of which is consistent with an octahedral μ^+ site with 4.9% dilation of the surrounding lattice atoms (later, by level-crossing resonance (LCR), nuclear electric quadrupole (NEQ) interaction in the nearest neighbor Cu was found to be constant, Luke *et al.*, 1991); (2) the hopping rate has a clear minimum at 30 K, above which the data can be fitted by an activation-type single exponential, consistent with the transverse-field data at high temperatures; (3) below 10 K and down to 0.5 K, the hopping rate increases

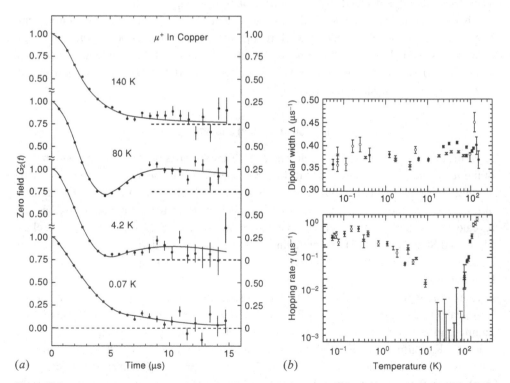

Figure 8.3 (*a*) A typical muon spin rotation/relaxation/resonance (μSR) time spectrum for the μ^+ in Cu and (*b*) temperature dependences of the dipolar widths and hopping rates extracted from the time spectra. Reproduced from Kadono *et al.*, 1989.

with decreasing temperature, as represented by $T^{-\alpha}$ with $\alpha = 0.67(3)$; (4) below 0.5 K, the hopping rate levels off at a value of around 0.5 μs^{-1}.

As shown later, the complete set of low-temperature data for the μ^+ hopping rate can be explained by applying the theory developed by Kondo (1984). This theory takes into account both muon–phonon and muon–electron interactions. By fitting to the experimental data, the parameter κ, representing the strength of the muon–electron interaction, was obtained to be 0.32. With respect to the leveling-off behavior below 0.5 K, it was also pointed out by Kondo that a broadening of the muon energy level of the order of 0.34 K (30 μeV) leads to precisely the kind of behavior observed (Kondo, 1984). This magnitude of level broadening can be expected as a result of residual impurities, dislocations, isotope mixture, and other factors.

Further experiments have been carried out to explore the details of the low-temperature behavior, employing the transverse field–muon spin rotation (TF-μSR) method. The following systematics were observed for the width of the transverse relaxation below 2 K: (1) the presence of a 95 p.p.m. Fe impurity significantly suppresses the decrease in width; (2) isotopically pure ^{63}Cu and ^{65}Cu samples yield almost the same width as does the natural Cu isotopic ratio. It was anticipated that there might have been some isotopic mixture effect as a result of strain–energy changes accompanying differences in zero-point motion (leading to volume differences between sites involving the majority isotope and those involving the 31% of ^{65}Cu present).

Similar types of measurements to explore μ^+ quantum diffusion phenomena have been conducted in Al (Karlsson *et al.*, 1995), as seen in Figure 8.4, and in Ta (Kadono *et al.*, 1997 a, b). In the case of Al, the effect of superconductivity on the diffusion rate has been

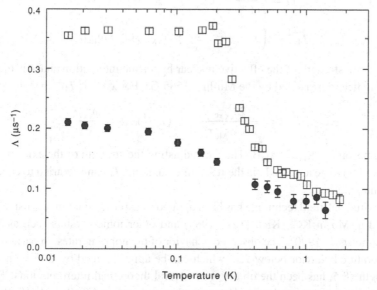

Figure 8.4 The relaxation rate reflecting μ^+ hopping in normal (filled symbols) and superconducting (open symbols) states of Al doped with 75 p.p.m. Li (Karlsson *et al.*, 1995).

learned for Al doped with 75 p.p.m. Li. As seen in Figure 8.4, a dramatic contrast in the μ^+ depolarization rate has been observed between the superconducting and nonsuperconducting states: the μ^+ diffusion rate is high in the superconducting state, but considerably lower in the normal state brought about by the application of a magnetic field. This enhanced μ^+ diffusion rate in the superconducting state had been theoretically predicted (Kagan and Prokof'ev, 1991); the diffusion rate is inversely proportional to the final state level broadening and, since the level broadening decreases with the gap energy, is thus an exponentially increasing function of the superconducting energy gap.

8.1.2 Mu diffusion in KCl and other ionic crystals

Mu, which is a neutral atom-like state composed of the μ^+ and an electron, undergoes a characteristic rapid diffusion through ionic crystals such as alkali halides or rare gas solids. Since Mu carries an electron, there are two types of magnetic interaction which lead to dipolar relaxation of the μ^+: (1) that between randomly oriented static nuclear dipoles and the Mu electron (nhf); and (2) that between the Mu electron and the μ^+ spin (Mu-hfs). Thus, the narrowing of the nuclear dipolar width due to the Mu motion which is sensed via the nhf interaction is amplified by the Mu-hfs before the μ^+ spin is depolarized. Developments have been made in the longitudinal relaxation method under applied longitudinal field to cover several orders of magnitude of the hop rate ($\nu = \tau_c^{-1}$; Kiefl et al., 1989).

By considering the transition rate between the Mu triplet spin states (ω_{12}) in the case where the hop rate (ν, τ_c^{-1}) is small compared with ω_{12}, the time dependence of the residual component of the Mu under longitudinal field shown in Chapter 6 becomes as follows:

$$p_z(t) = \frac{1 + 2x^2}{2(1 + x^2)} e^{-t/T_1}$$

$$T_1^{-1} \cong \left(1 - \frac{x}{\sqrt{1 + x^2}}\right) 2\delta_{ex}^2 \tau_c / (1 + \omega_{12}^2 \tau_c^2)$$

where δ_{ex} is the strength of the effective nuclear hyperfine interaction and x is the external longitudinal field normalized by the Mu-hfs (1585 G). For $x \ll 1$, T_1^{-1} becomes:

$$T_1^{-1} = \frac{2\delta_{ex}^2 \tau_c}{1 + \omega_{Mu}^2 \tau_c^2}, \qquad (T_1^{-1})_{max} = \frac{\delta_{ex}^2}{\omega_{Mu}}$$

where ω_{12} becomes $\omega_{Mu} = \gamma_{Mu} B$. Thus, by adjusting the strength of the external field, the range of $\tau_c (\nu^{-1})$ to be accessed via the μSR time range for T_1 can be varied to cover several orders of magnitude.

The longitudinal-field technique has been applied to explore quantum diffusion phenomena related to Mu in KCl (Kiefl et al., 1989) and other ionic crystals such as NaCl and KBr (Kadono et al., 1990). The observed behavior of the hopping rates, in particular, of the low-temperature behavior below T_{min} which can be approximated by $T^{-\alpha}$ with $\alpha \approx 3$, as seen in Figure 8.5, has been the objective of several theoretical interpretations. The low-T behavior was first interpreted as evidence of coherent hopping (Kiefl et al., 1989), and was later explained using two-phonon scattering theory (Kagan and Prokof'ev, 1990).

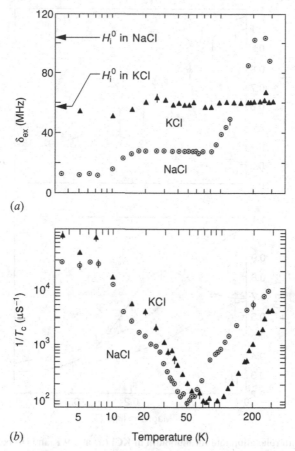

Figure 8.5 Temperature dependence of (a) the nuclear hfs parameter δ_{ex} (with arrows indicating values estimated for a hydrogen atom case) and (b) hopping rates of Mu in KCl and NaCl observed by the high-longitudinal-field relaxation method (Kadono *et al.*, 1990).

When one enters into the temperature region where $T \ll \Delta$ (an effective tunneling matrix element), the Mu atom falls into a coherent Bloch state; the Mu begins to occupy a well-defined eigenstate of energy (i.e., the bottom of the Mu energy band) which leads to the delocalization of the state vector due to Heisenberg's uncertainty principle. There, the longitudinal muon spin relaxation rate induced by fluctuation of local magnetic fields acting on the Mu orbital electron may strongly reflect the shape of the Mu density of states $\rho(\varepsilon_k)$. In particular, one would expect a strong modulation of the relaxation rate $(1/T_1)$ when the Zeeman frequency coincides with van Hove singularities (Kondo, 1999). The experimental results of Mu spin relaxation are shown in Figure 8.6, where one can notice a clear difference between the data at 3.9 K and below 10 mK (Kadono *et al.*, 1999). The spectral density below 10 mK is reproduced by assuming a Lorentzian distribution plus a constant background relaxation ($\cong 2.5 \times 10^{-5} s^{-1}$), as shown by the solid curve. This peak structure in the spectral density below 10 mK is a clear signature that the muonium is in the Bloch state.

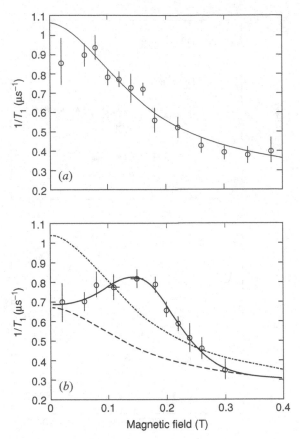

Figure 8.6 Muon spin relaxation rate for muonium in KCl (*a*) at 3.9 K and (*b*) below 10 mK. Dot-dashed curve in (*b*) is the best fit to a Lorentzian spectrum, whereas the dashed curve is the Lorentzian spectral component fitted in conjunction with a Gaussian peak around 0.15 T to give the solid curve.

8.2 Probing Mu/μ^+ center in semiconductors and insulators

8.2.1 Methods so far applied

The signals from muonium-like paramagnetic states in solids can be obtained using various types of experimental arrangement: the spin rotation signal in a weak transverse field ($\sim$ a few G) is useful to show the *existence* of the Mu-like state, while the two-frequency precession signals seen in medium transverse fields ($\sim$100 G), as seen in Figure 1.3, can be used to determine the hyperfine coupling constant. Several other methods also exist. When the Mu-like state is formed in solids, it is usually under the influence of various perturbing local magnetic fields such as nuclear dipolar fields and atomic dipolar fields from the surrounding paramagnetic moments. Since these perturbing magnetic fields can easily destroy the phase-coherent precession of the muonium spin, it is sometimes very difficult to obtain a muonium signal using these conventional TF rotation methods.

When a high enough field is applied to the Mu-like paramagnetic state (large compared to the hyperfine field ν_0 from the bound electron on the μ^+), the precession frequency (ν_{12} or ν_{34}) becomes weakly dependent on the longitudinal field and becomes insensitive to static inhomogeneous fields. Thus, the Mu/radical spin rotation can be seen even in the presence of the perturbing fields, given a detection system with a high enough time resolution. This method has been developed and used in experiments at PSI for studies of radical states in chemical substances, as described later. For Mu-like states with ν_0 close to the value in vacuum, making use of high fields in combination with high-time-resolution μSR spectrometry (Kiefl et al., 1984), the Mu states were successfully observed in alkali halides (Figure 8.7) and other systems, as summarized by Cox and Symons (1986). In some cases, observations had not hitherto been possible due to the strong perturbing fields from the surrounding nuclear moments.

When a strong longitudinal magnetic field is applied along the muon polarization direction, the μ^+ polarization can be restored to its full value by decoupling the perturbing effects due to static and/or dynamic nuclear and/or electronic fields. For the actual identification of the states of the muon (e.g., whether the muon is in a paramagnetic Mu-like state or diamagnetic μ^+ state), the muon spin r.f. resonance method is of considerable use. By adjusting the frequency of the r.f. field to the Zeeman frequency of Mu or μ^+, we can clearly identify which states are present. In the case of Mu resonance, there are several schemes depending upon the associated energy levels in the Breit–Rabi formula (see Figure 1.3):

1. High-frequency: $f = \nu_{14}$ or ν_{34}, under a weak longitudinal field. This high-frequency microwave resonance ($f = 4.4\,\text{GHz}$ for free muonium) is used for precise determination of the hyperfine coupling constant of muonium in various inert gas systems.
2. Low-frequency: ν_{12}, ν_{23} resonance. This corresponds to the $F = 1$ precession in low-field TF-μSR, but it will reveal a resonance signal even in the presence of small nuclear field or in the case that the muonium is formed via a precursor state.
3. Intermediate-frequency: ν_{12}, ν_{34} resonance at high field. This technique is particularly useful, since the applied field serves to decouple local perturbing fields.

8.2.2 Muonium-like states in semiconductors

This subject has been studied extensively since the birth of μSR. Review articles (Patterson, 1988; Kiefl and Estle, 1990) give comprehensive coverage of the historical developments in μSR studies on semiconductors.

In addition to the diamagnetic μ^+ state, there are two basic muonium-like states known as normal Mu and anomalous Mu (Mu*), as seen in historical μSR data in Si at 77 K (Brewer et al., 1973), shown in Figure 8.8; normal Mu has a binding energy slightly weak compared to that of Mu in vacuum, while anomalous Mu has a binding energy an order of magnitude smaller than Mu in vacuum. In Si, the system which has been most extensively studied thus far, the fraction of each state changes with temperature and with dopant concentration of either p-type or n-type, as summarized in Figure 8.8. Generally speaking, as the temperature becomes lower and the dopant concentration becomes smaller (below 10^{18}/cc), both normal

Table 8.1 Summary of muonium-like centers in various
semiconductors

| | Mu | | Mu* | | |
Host	A^{μ}(MHz)	η_s^2(Mu)[a]	$A_{\parallel}$(MHz)	$A_{\perp}$(MHz)	η_s^2(Mu*)[b]
C	3711(21)[c]	0.831	+167.98(6)	−392.59(6)[e]	−0.0461
Si	2006(2)[c]	0.449	−16.82(1)	−92.59(5)[e]	−0.0151
Ge	2359.5(2)[c]	0.529	−27.27(1)	−131.04(3)[e]	−0.0216
GaP	2914(5)[d]	0.653	+219.0(2)	+79.48(7)[d]	+0.0286
GaAs	2883.6(3)[d]	0.646	+218.54(3)	+87.87(5)[f]	+0.0294

[a] η_s(Mu): A^{μ}/A_{free}(A_{free}: 4463.302MHz).
[b] η_s(Mu*): $1/3(A_{\parallel} + 2A_{\perp})/A_s^{\text{free}}$.
[c] Holzshuh, E. *et al.* (1983). *Phys. Rev.* **B27**, 102.
[d] Kiefl, R.F. *et al.* (1985). *Phys. Rev.* **B32**, 530.
[e] Blazey, K.W. *et al.* (1983). *Phys. Rev.* **B27**, 15.
[f] Kiefl, R.F. *et al.* (1987). *Phys. Rev. Lett.* **B34**, 1474.

Mu and anomalous Mu become increasingly stable. Note that the missing Mu signal at elevated temperatures is mainly due to Mu reactions in the solid.

The hyperfine coupling constants of the normal Mu states at the lowest temperature in all the semiconductors so far studied are summarized in Table 8.1. The values do not change monotonically for the series C, Si, Ge. Structurally speaking, normal muonium should be considered as an isotropic donor located at an interstitial site. The structure and hyperfine coupling of the normal muonium state have been the objects of a series of theoretical studies using electronic structure methods such as the first-principle Hartree–Fock cluster theory (Sahoo *et al.*, 1985) and band structure method based upon the density functional theory (Van de Walle *et al.*, 1988).

The corresponding values for anomalous Mu (Mu*), in Si as well as in other representative semiconductors, are summarized in Table 8.1. The hyperfine fields for these Mu states were experimentally observed to be anisotropic, and so the overall angular momentum of these states must be larger than one. Experimental studies of both the TF-μSR and the LCR, as seen in Figure 8.8, have revealed that anomalous Mu states are located at the bond-center site. Most of the theoretical calculations predict that the global minimum in the potential-energy surface of neutral Mu centers in Si is Mu*, which is expected at the bond-center site, while normal Mu center at the interstitial site (Mu_T^0) is expected to be a metastable state appearing in the μSR spectrum at low temperature with the largest fractions. The M_T^0 center is observed to be rapidly converted into Mu_{BC}^0 under illumination (Kadono *et al.*, 1994).

8.2.3 Muonium in alkali halides

After the development of the high transverse field–muon spin rotation (HTF-μSR) method, a series of measurements were undertaken on the hyperfine coupling constants of Mu-like

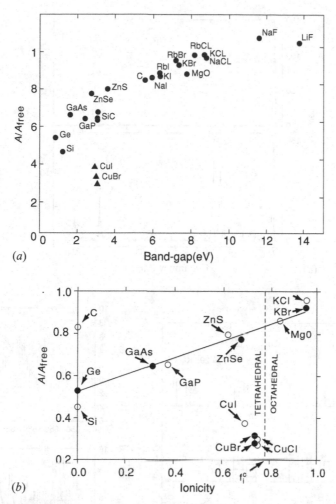

Figure 8.7 Summary of the hyperfine coupling constants for Mu in various semiconductors and insulators represented as a function of band gap (Cox and Symons, 1986) and ionicity (Kiefl *et al.*, 1986).

states in alkali halides. The results of these are summarized in Figure 8.7. Qualitatively, as is shown in the figure, the hyperfine coupling constants have a monotonic dependence on the band gap energy (or the lattice constant of the host).

The Mu state in alkali halides is unstable, particularly at elevated temperatures (above 200 K). But even at the lowest temperature, it was found by the resonance method that the Mu state is converted into a diamagnetic state as a result of a chemical reaction in the solid (Morozumi *et al.*, 1986). Also, in order to account for a large missing fraction (around 50%) observed at the lowest temperatures, we should consider the possibility that there is a precursor Mu-like state from which the other unobserved Mu-like states are formed.

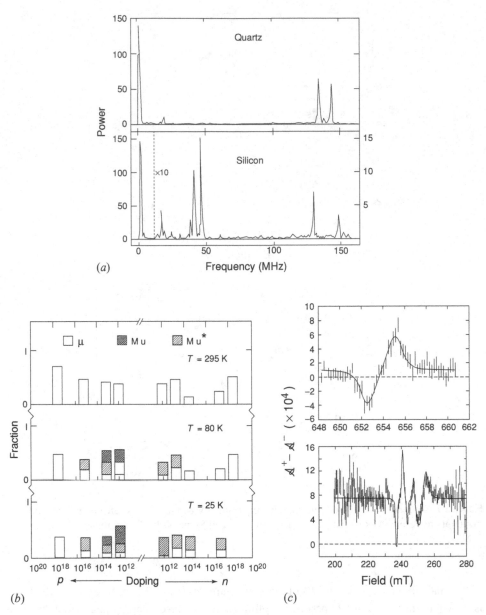

Figure 8.8 (*a*) Transverse field-muon spin rotation (TF-μSR) data at 10 mT in quartz at room temperature and [111] Si at 77 K (Brewer *et al.*, 1973). (*b*) Fractions of the μ^+ states measured in intermediate-field TF-μSR in typical semiconductors with various dopants and at various temperatures (Patterson, 1988). (*c*) Level-crossing resonance spectra for Mu* in Si with the longitudinal field along <110> axis (Kiefl *et al.*, 1988).

8.3 Muonium radicals in chemical compounds

The Mu radical spectroscopy has been started by observing various spin rotation lines under relatively high magnetic field of 0.1 T ~ T (Roduner *et al.*, 1978, 1981). Later, significant progress has been marked by employing LCR spectroscopy (Kiefl *et al.*, 1986).

The Mu radical has a significant difference from the analogous H radical in the following sense:

1. Mass is different ($M_{Mu} \approx \frac{1}{9} M_H$), so that associated electronic structure as well as vibrational/rotational spectrum is different due to a mass correction in the same way as deuterated or tritium labeled radicals.
2. There are no interactions among radical species (no interaction among Mu radicals) so that energy transfer due to a matching of the energy levels under zero external field is strictly forbidden.
3. Since Mu is formed by injecting a high-energy μ^+, one must expect a formation of some excited states of the Mu radical at the time of formation.

Since the details of the present research activities in this field are beyond the scope of the present book, it is suggested that the reader refer to summary reports (Roduner, 1988, 1999).

8.4 Probing electron transfer in polymers and macromolecules: labeled-electron method

The spin dynamics of paramagnetic conduction electrons in macromolecules such as conducting polymers can directly reflect the nature of electron conductivity. The role of μSR (with the probe species being μ^+) is significant in this field, as explained by the following scenario (Figure 8.9).

During the slowing-down process in soft materials such as conducting polymers, the injected μ^+ picks up one electron to form a neutral atomic state, muonium. Muonium is then thermalized, and bonds chemically to a reactive site on the molecule. Then, depending on the nature of the molecule, the electron brought in by the μ^+ can exhibit several characteristic behaviors, including localization to form a radical state and/or a linear motion along the molecular chain. These behaviors can be detected with high sensitivity by measuring the spin relaxation process of the μ^+ using the μSR method; muon spin relaxation occurs in this case through a magnetic interaction between the μ^+ and the localized and/or moving electron produced by the μ^+ itself – the labeled-electron method.

The most significant observations in these μSR studies can be summarized as follows. In longitudinal relaxation measurements, due to the nature of the interaction between the moving electrons and the stationary muons, the characteristic dimensionality of the electron motion can be studied by varying the externally applied magnetic field (B_{ext}) and observing the dependence of the muon spin relaxation rate (λ_μ); for one-dimensional electron motion $\lambda_\mu \propto (B_{ext})^{-1/2}$, for two-dimensional electron motion $\lambda_\mu \propto (\alpha - \beta \log B_{ext})$, where α and β are constants, and for three-dimensional electron motion λ_μ does not usually have significant B_{ext} dependence (Butler *et al.*, 1976).

Progress has been made in the theoretical understanding of this paramagnetic relaxation process by Risch and Kehr, who considered the direct stochastic treatment of the

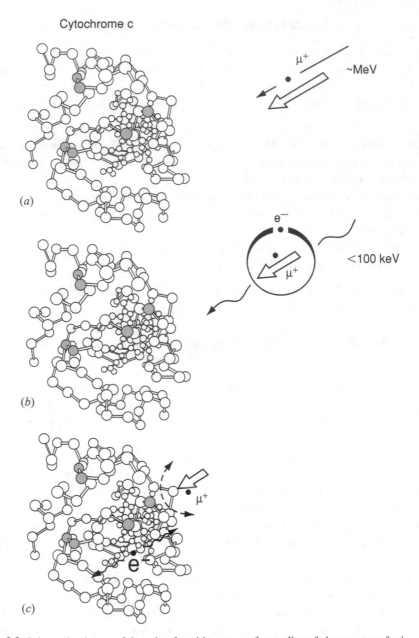

Figure 8.9 Schematic picture of the role of positive muons for studies of electron transfer in macromolecules. (a) Energetic μ^+ introduction, (b) a process of electron pick-up to form Mu during μ^+ slowing-down, and (c) thermalization and physical/chemical bonding of the μ^+ and/or Mu and release of the brought-in electron.

random-walk process of a spin which is rapidly diffusing along a topologically one-dimensional chain (Risch and Kehr, 1992). An error-function-type longitudinal relaxation function (hereafter called the R-K function) $G(t) = \exp(\Gamma t)\,\text{erfc}(\Gamma t)^{1/2}$ was proposed for $\lambda t_{max} \gg 1$, where λ is the electron spin flip rate, t_{max} the experimental time scale, and Γ a relaxation parameter. In this theoretical treatment, in the case of topologically one-dimensional electron motion, Γ is proportional to $1/B_{ext}$.

Note that the introduction of an electron in the form of energetic Mu is, at this moment, an assumption. Since the radiolysis effect is known to be another major mechanism for electron pickup (see Chapter 3), one should check experimentally the validity of the assumption by applying an electric field. Another possible role of the μ^+ introduction into the biological macromolecule is participation in a biological process such as the H^+ pumping mechanism. Once known to be effective these mechanisms may open up other possible life science studies with muons.

8.4.1 Formation and decay of muonic radicals in conducting polymers

This idea of the sensitive detection of the electron behavior in macromolecules using muons has been successfully applied in studies of electron transport in conducting polymers. A soliton-like motion of the μ^+-produced electron in *trans*-polyacetylene has been observed, which contrasts with the localization seen in *cis*-polyacetylene following the formation of a radical state (Nagamine *et al.*, 1984; Ishida *et al.*, 1985).

Let us describe first the basic properties of polyacetylene, which is the simplest polymer based upon CH- or CD- chains with alternating double and single bonds, as is shown in Figure 8.10. Such a system is said to be *conjugated*. It is known to be an organic semiconductor. It has two isomers, *trans*- and *cis*-polyacetylene. Experimental studies progressed tremendously after the invention of new techniques to produce stable thin-film samples (Shirakawa and Ikeda, 1971). The ground-state *trans*-polyacetylene can be formed through thermal isomerization by heating the *cis*-isomer above 150° C. In this isomerization process, it is known from the electron spin resonance (ESR) experiments that an unpaired electron is created in about one chain out of 10. As shown in Figure 8.10, the unpaired electron is associated with a change in bond alternation at the electron's location. Theoretical work by Su *et al.* (1979) predicted that the one-dimensional motion of this unpaired electron along the chain in the *trans*-isomer should take the form of a "soliton," like the nonlinear dynamic motion of a chain of coupled pendulums. This "soliton" motion exists only in the *trans*-isomer because of the degeneracy of the two bond-alternated states. The same theory also predicted that the wave function of the unpaired electron should be distributed over many carbon sites (up to 20). Many experiments using ESR, nuclear magnetic resonance (NMR), and other techniques have studied the behavior of the unpaired electron in polyacetylene; almost all of the results are consistent with the soliton picture.

In μSR measurements, μ^+ were injected into *cis*- and *trans*-$(CH)_x$ and $(CD)_x$, and the time spectrum measured under various longitudinal fields at room temperature, as shown in Figure 8.10 (Nagamine *et al.*, 1984). A clear contrast can be seen in the μ^+ polarization phenomena between the *cis*- and *trans*-isomers; in *cis*-$(CH)_x$, the μ^+ polarization at $t = 0$ increases steeply with increasing applied field, with no significant relaxation observed at

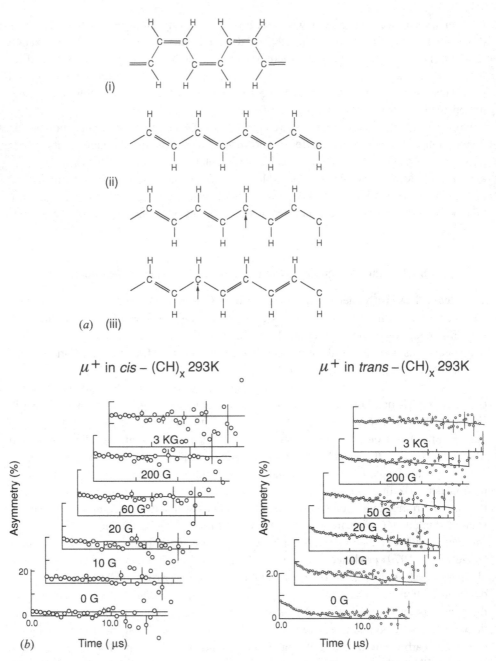

Figure 8.10 (*a*) Chemical structure of *cis*-(i) and *trans*-(ii) polyacetylene and soliton motion of a paramagnetic electron along the chain of *trans*-polyacetylene (iii). (*b*) The muon spin rotation/relaxation/resonance (μSR) time spectrum for the μ^+ in *cis*- and *trans*-polyacetylene at room temperature under various applied longitudinal fields.

any field, while in *trans*-$(CH)_x$ the initial polarization remains almost constant at its full value with significant field-dependent changes in relaxation.

The observed spin polarization behavior in *cis*-$(CH)_x$ shown in Figure 8.10 is reminiscent of the decoupling or quenching pattern of the μ^+ polarization observed when a muonium-like paramagnetic state is formed. In some systems, this state is known as a muonium-substituted radical. The structure of the radical state in *cis*-polyacetylene was further studied using the high transverse field rotation method (Nishiyama *et al.*, 1986). In these measurements, the radical state formed by the μ^+ in *cis*-polyacetylene was found to have a hyperfine coupling constant corresponding to an electron localization at a radius three times larger than in free muonium. The hyperfine interval was determined to be $A_0 = 91.0(2)$ MHz, compared to the 300 MHz expected for a fully localized unpaired π electron. According to the present understanding of muonic radical formation in various chemical substances, the following picture seems to be adequate: the μ^+ picks up one electron to form epithermal muonium during its slowing-down process, and this muonium is then attached to a carbon of the double bond of polyacetylene, with the unpaired π-electron of the double bond being localized at several carbon sites distributed nearby forming a radical electron.

By contrast, the μ^+ polarization phenomena in *trans*-$(CH)_x$ are reminiscent of the picture in which the diamagnetic μ^+ is subject to spin relaxation due to a dynamically perturbing field. As a natural explanation based upon the "soliton" model, the following picture was proposed (Figure 8.11). The μ^+ interacts with a "soliton" and is subject to spin relaxation due to a rapidly fluctuating magnetic field from the moving electron. It should be noted that the soliton exists only on one side of the μ^+ location along the chain. The picture of the μ^+ relaxation as being due to a one-dimensionally moving spin is supported by the fact that the μ^+ relaxation rate (a single exponential-type relaxation is taken here) follows an inverse-root dependence on the applied field, as demonstrated in Figure 8.11.

Let us compare the μ^+ relaxation rate to that of ^1H obtained by NMR, which is around $2.4 \times 10^3 H^{-1/2}s^{-1}$ at room temperature. The observed $T_1(\mu)^{-1}$ for the μ^+ in both $(CH)_x$ and $(CD)_x$ is two orders of magnitude larger than $T_1(p)^{-1}$ at the same external field. However, the ratio of $T_1(\mu)^{-1}$ and $T_1(p)^{-1}$ should be scaled by the square of the gyromagnetic ratio, γ_μ^2/γ_p^2, which is nearly 10. Even after this correction there still remains a difference of around a factor of 10 in the ratio of T_1^{-1}. This fact supports the proposed picture of the labeled-electron method; the μ^+ probes the soliton produced by itself in its own chain, while in the NMR case the soliton is that produced in thermal isomerization, with a density of one per 10 chains.

In order to study the diffusion properties of the muon-produced soliton, the temperature dependence of the spin relaxation was measured (Ishida *et al.*, 1985). This result indicates a model in which the source of the relaxation, namely the soliton, disappears from the chain above 281 K by interchain diffusion with time after it is produced by the muon's arrival (Ishida *et al.*, 1985).

In the succeeding experiment on polyaniline, the usefulness of the R-K function was confirmed experimentally for the polaronic motion of conduction electrons in polyaniline, and a method for obtaining 1D- and 3D-diffusion rates from the fitted R-K functions was pointed out (Pratt *et al.*, 1997).

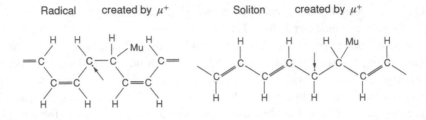

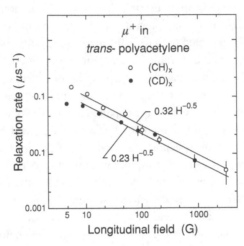

Figure 8.11 (*a*) The μ^+-induced soliton in *trans*-polyacetylene. (*b*) An example of $\sqrt{H}$ dependence – the field dependence of the μ^+ relaxation rate in *trans*-polyacetylene at room temperature.

8.4.2 Probing electron transfer in biological macromolecules; μSR life science

Electron transfer in macromolecules such as proteins is an important mechanism for various biological phenomena. A number of experimental investigations have been carried out using a variety of techniques to explore the electron transfer phenomena in proteins and related chemical compounds. However, almost all the existing information on the electron transfer process has been obtained by essentially macroscopic methods, which measure the evolution of the electron transfer from donor to acceptor.

In order to obtain microscopic information on electron transfer in protein macro-molecules, the μSR method, as described in the previous subsection, offers great potential. Depending on the nature of the molecule, the electron accompanying the μ^+ into the molecule can have a variety of characteristic behaviors, including localization to form a radical state and linear motion along the molecular chain. These behaviors can be distinguished with high sensitivity by measuring the spin relaxation process of the μ^+ using the μSR method.

Of the very large number of proteins in existence, cytochrome c is one which has attracted much attention, since it plays an essential role in the respiratory electron transport chain in mitochondria; it holds a position next to the site of the final process of the cycle and transfers electrons to the surrounding oxidase complex. Experiments on the μ^+ relaxation in cytochrome c were conducted using an intense pulsed beam of 4 MeV μ^+ at Institute of Physical and Chemical Research–Rutherford Appleton Laboratory branch (RIKEN-RAL) (Nagamine et al., 2000).

At each of the measurement temperatures, the μ^+ relaxation function was found to have an external field dependence (Figure 8.12). The observed relaxation functions $G(t)$ were fitted with the R-K function, whereupon the longitudinal relaxation parameter Γ obtained at various temperatures was found to decrease monotonically with increasing B_{ext}. On closer inspection of the B_{ext} dependence of Γ, there seem to be two separate field regions exhibiting different behavior: (1) a region of weak field dependence (lower field); and (2) a $(B_{ext})^{-1}$ dependence region (higher field; Figure 8.12). The latter region exhibits the characteristic μ^+ spin relaxation behavior due to the linear motion of a paramagnetic electron. The critical value of the cut-off field where the second type of behavior supersedes the first has a significant temperature dependence: it grows smaller with decreasing temperature. It can further be seen that the temperature dependence of the cut-off field can be represented by the sum of two activated components of the form $\exp(-E_a/kT)$, where E_a is an activation energy; of these, one has an activation energy of 150 meV (dominant above 200 K), while the other has an activation energy of less than 2 meV (dominant below 200 K). In the context of a protein such as cytochrome c with coils and folds in its structure, the "interchain" diffusion might perhaps be interpreted as "interloop" jumps, which could well be strongly activated by the increased thermal displacement of the protein chain occurring above the glass transition temperature of 200 K.

The most important unknown factors in the present μ^+SR studies are the distribution of locations of the μ^+ bonding sites, together with the corresponding uncertainty in the electronic structure associated with μ^+ and its environment and the site from where the electron commences its linear motion. For the purpose of elucidating these matters, muon spin r.f. resonance and LCR will be the most helpful techniques. Theoretical studies on the possible μ^+ and Mu sites in cytochrome c have been carried out using the Hartree–Fock procedure, suggesting the nitrogen of the pyrrole ring and/or negatively charged parts of amino acids (Cammarere et al., 2000; Scheicher et al., 2001; 2002).

Experiments have been extended to other proteins, such as: (1) myoglobin, which is known to be important in oxygen transport with a molecular structure similar to that of the electron-transfer protein; and (2) cytochrome c oxidase, which is known to stay at the terminal position of electron transfer in the mitochondria aspiration cycle. A characteristic difference in temperature dependence in interchain diffusion rate was seen between cytochrome c and myoglobin, suggesting a difference between "natural" and "artificial" electron transfer in protein (Figure 8.13). Also, electron transfer is limited at room temperature, while enhanced transfer at low temperature was seen in cytochrome c oxidase, suggesting possibilities for the exploration of the electron transfer path.

The labeled electron method was also applied to DNA. Electron-transfer phenomena in DNA are known to be important not only for understanding damage and repair mechanisms, but also for possible applications to new biodevices. The experimental finding

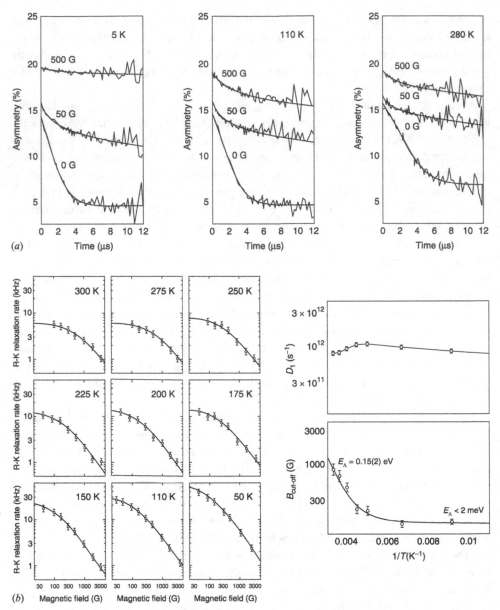

Figure 8.12 (*a*) Typical μ^+ spin relaxation time spectra in cytochrome c at 5 K, 110 K, and 280 K under external longitudinal fields of 0 G, 50 G, and 500 G. For finite field the curves show best fits using the R-K function. (*b*) The R-K relaxation parameter Γ versus external longitudinal magnetic field for the μ^+ in cytochrome c at various temperatures. The $(B_{ext})^{-1}$ dependence part can be seen to become significant in the higher field region, and the critical field (cut-off field) for the onset of the $(B_{ext})^{-1}$ dependence can be seen to have a clear temperature dependence.

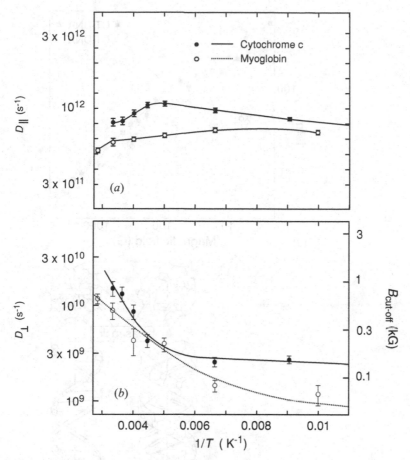

Figure 8.13 (*a*) Temperature dependence of the parallel diffusion rate of an electron in cytochrome c and myoglobin derived from the B^{-1}-dependent part of the relaxation curve. (*b*) Perpendicular diffusion rate, derived from the cut-off field determined in the Γ versus B data, plotted against the inverse temperature.

(Meggers *et al.*, 1998) of a possible electron transfer between G (guanine) bases has accelerated both experimental and theoretical studies. The Yamanashi U–Institute of Physical and Chemical Research (RIKEN)–High Energy Accelerator Research Organization (KEK)–Oxford–Tokyo Kasei Gakuen U–Juelich collaboration has successfully conducted a μSR experiment at RIKEN-RAL on an oriented DNA sample in both A and B conformations of the DNA extracted from calf thymus (Figure 8.14), and observed electron transfer in DNA at room temperature (Torikai *et al.*, 2001). There, after analyzing μSR data along the procedure mentioned above, the relaxation parameters Γ were found to take an inverse-field dependence in both A and B cases above 80 G (Figure 8.14), suggesting an existence of quasi-1D rapid diffusion of electrons in DNA. However, there is a clear difference in the field dependence below 80 G – a continuation of the inverse-field dependence in the A case, but no field dependence in the B case. The result demonstrates that some aspects of the quasi-1D

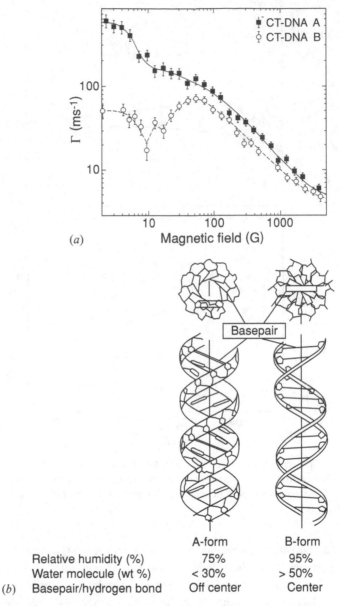

Figure 8.14 Field dependence of the observed relaxation parameters in (a) both A- and B-form of the DNA and (b) crystalline structure of DNA in A-form (left) and B-form (right) with a difference in relative humidity, water molecule concentration (wt %) and basepair/hydrogen bond.

diffusion seen in the labeled-electron method correlate with the arrangements of base pairs. The result is somewhat consistent with a picture of electron hopping through base pairs.

The μ^+SR method, as described above, where the high efficiency of the technique should be emphasized, can easily be extended to the study of proteins or DNA in various chemical and biological environments. Most importantly, because of the initial high-energy nature of the probe, this method can be applied to proteins or DNA in vivo. As one example for the future, it may be possible to use the technique to obtain new information on the basic functions of brain activity.

8.5 Muonium chemical reaction

The μSR method can be used as a sensitive probe of chemical reactions in which the muon charge/spin state changes from paramagnetic Mu to diamagnetic μ^+ or vice versa. A clear change in the response to an applied external field (from 104 μ_μ to 1 μ_μ) is the main clue to be used in distinguishing Mu from μ^+.

Detailed descriptions of the knowledge obtained from studies of such Mu chemical reactions can be found in a review paper by Fleming *et al.*, (1992). Here, some of the most significant developments will be summarized, with emphasis given to the points of uniqueness of Mu chemical reaction studies compared to conventional chemical reaction studies. Some important features of these studies can be summarized as follows: (1) a drastic isotope effect compared to H, D, and T in hydrogen reactions; (2) a unique time range for reaction rate determination; and (3) some possibility of final state identification by selecting a diamagnetic state.

8.5.1 Mu chemical reactions in gases

Hydrogen atom reaction kinetics in simple gases is one of the most fundamental topics of physical chemistry. Consequently, it is one of the first priorities in Mu atom chemistry to learn the rate of the standard reaction $H + Cl \rightarrow HCl$ with H isotopically substituted by Mu. By changing the temperature and the kinetic energy of Mu, one can experimentally obtain the potential surface on which the chemical reaction occurs. Experiments of this kind have been in progress since the early 1970s. The change in vibrational zero-point energy (in the transition state and/or product) is now known to be one of the most important contributions to the origin of the characteristic difference between H and Mu reaction rates.

8.5.2 Mu chemical reactions in aqueous solutions

In the early stages of Mu chemical reaction studies, again due to the zero-point energy effects on the initial, final, or transition states along the reaction coordinate resulting from the muon mass effect and the through-barrier tunneling effects, it was found that the Mu reaction rate is a sensitive tool to test reaction rate theories (Percival, 1980). Since then, along with spectroscopic studies of radical states formed by the introduction of Mu or μ^+ in organic liquids, Mu rate-constant studies in liquid-phase chemistry have been extended.

8.5.3 Mu chemical reactions in solids

By the use of advanced μSR methods such as muon spin r.f. resonance, one can learn about the final states of chemical reactions. Such studies have been used to explore Mu to diamagnetic μ^+ chemical reactions both in alkali halides and in semiconductors.

"Normal" Mu located at the interstitial site, formed in Si directly upon μ^+ introduction, was found to undergo ionization to μ^+ at room temperature, by using the muon spin resonance method to identify the final state (Nishiyama, 1992; Kreitzman et al., 1995). Similarly, as described earlier (section 8.2.2) at very low temperature under photon irradiation metastable Mu trapped at the interstitial site in Si after muon implantation undergoes a stabilization reaction towards its lowest energy state, namely ionized μ^+ at the bond-centered site (Kadono et al., 1994).

Mu chemical reactions were also found in various alkali halides through measurements detecting final μ^+ states formed from the initial paramagnetic Mu state, itself identified by the muon spin resonance method as a recovered state in longitudinal field repolarization measurements. Systematic studies of the activation energies have been given for Mu in KCl, NaCl, KI, and other systems (Nishiyama et al., 1985; Morozumi et al., 1986).

8.5.4 Mu chemical reactions on solid surfaces

Some insulators such as SiO_2 or Al_2O_3, in powdered form, can serve as suitable materials for the production of thermal Mu. Thus, by placing a target consisting of powders of these materials in the muon beam path at low temperatures, one can expect the formation of Mu, which can then migrate to the surface of the powdered solid. If some atom or molecule is then introduced as a reaction partner, one can make measurements of the chemical reaction rates of Mu on the surface. Thus the reaction rate with O_2 was measured for Mu on the SiO_2 surface in the temperature range of 20–100 K (Kempton et al., 1990).

8.6 Paramagnetic μ^-O probe

As described in Chapters 4 and 6, when injected into condensed matter, the μ^- creates a spatially extended $(Z - 1)$ nucleus. The μ^-SR (with μ^-O as probe) and enriched ^{17}O NMR can give direct information regarding the O-site in, e.g., high-T_c superconductors. While ^{17}O NMR requires isotopically enriched samples, the μ^-O method can be applied to a standard sample with ^{16}O nuclei, thus avoiding any possible perturbation which might accompany isotopic substitution. Thus, the main advantage over the NMR experiment is an easiness of access to single crystal samples. Also, the fact that μ^-O is a purely magnetic probe without a quadrupole moment makes the interpretation of the experimental results much simpler compared to ^{17}O data.

The other unique feature of the $(\mu^-$O) probe is the rearrangement in the local electronic structure to accommodate the exotic $(\mu^-$O) core. When a muon is captured by a nucleus of charge Z, a large number of electrons are ejected from the μ^-Z atom during the earlier stages of the muonic cascade where the Auger effect is dominant. Either during or after the end of the cascade, the μ^-Z system will be repopulated by the surrounding electrons. In most solids, particularly in metals, the time required for repopulation is short enough compared to that for the muonic cascade $(10^{-14}-10^{-12}$ s) that the rearrangements of electrons can be completed without causing any additional depolarization. Thus, the electronic state corresponding to the $(Z - 1)$ quasinucleus is formed before the start of the μ^-SR measurement for the polarized ground state of (μ^-O) atoms. However, in insulators or semiconductors like high-T_c superconductors, with less mobile electrons, the screening of one nuclear charge by the muon is compensated for by the loss of an electron from the valence shell around the $(Z - 1)$ nucleus of the (μ^-Z) atom leading to local charge neutrality, which produces a paramagnetic hole state.

Temperature and crystalline axis dependence were measured for (μ^-O) in high-T_c superconductor $La_{2-x}Sr_xCuO_4$ ($x = 0.1$). Two different (μ^-O) signals were observed: one (site A) has a small shift (less than 0.3%) and is almost independent of temperature and crystalline axis, while the other (site B) has a large shift (up to 4%) and is strongly dependent on both temperature and crystalline axis (Torikai $et\ al.$, 1993). The theoretical understanding of the observed paramagnetic shifts and the determination of the location of the (μ^-O) state has been obtained by carrying out Hartree–Fock calculations on the structures corresponding to both the possible oxygen sites (apical and planar) (T. P. Das, private communication, see also Srinivas $et\ al.$, 1997). The results indicate that in both cases the μ^-O system in which an electron has been lost from O^{2-} by an Auger process is more stable than the μ^-O^{2-} system, the stability being stronger in the apical case. Also, for the apical case the hyperfine field is substantially stronger for the μ^-O system as compared to μ^-O^{2-}, the difference being much smaller for the planar case. Further, the results suggest that the anisotropic Knight-shift data are associated with the apical system while the planar system leads to less anisotropic data. Thus, the paramagnetic shift data (site B) can be considered to be evidence for the paramagnetic (μ^-O) state at the apical site.

The "paramagnetic" (μ^-O) system associated with the apical oxygen site is thus a new and important probe for understanding the nature of the superconducting current in high-T_c superconductors; a characteristic relaxation pattern of the paramagnetic electron can be expected to occur due to interaction with the supercurrent, and this relaxation can be detected via the (μ^-O) probe. An experiment aiming to observe the effect of interaction between the paramagnetic electron weakly bound to the apical (μ^-O) state and the supercurrent in the high-T_c superconductor LaSrCuO was conducted by measuring the spin relaxation of the bound μ^- in (μ^-O) under zero applied field (Torikai $et\ al.$, 1994). In ZF μSR, any change in spin relaxation rate on passing through T_c cannot be due to macroscopic static fields but is sure to be of dynamic origin.

REFERENCES

Brewer, J.H. $et\ al.$ (1973). $Phys.\ Rev.\ Lett.$, **31**, 143.

Butler, M.A. $et\ al.$ (1976). $J.\ Chem.\ Phys.$, **64**, 3592.

Cammarere, D. $et\ al.$ (2000). $Physica\ B$, **289–90**, 636.

Clawson, C.W. $et\ al.$ (1983). $Phys.\ Rev.\ Lett.$, **51**, 114.

Cox, S.F.J. and Symons, M.C.R. (1986). $Chem.\ Phys.\ Lett.$, **126**, 516.

Fleming, D.G. $et\ al.$ (1992). In $Perspectives\ of\ Meson\ Science$, ed. T. Yamazaki, K. Nakai, and K. Nagamine, p. 219. Amsterdam: North Holland.

Flynn, C.P. and Stoneham, A.M. (1972). $Phys.\ Rev.$, **B1**, 3966.

Gurevich, I.I. $et\ al.$ (1972). $Phys.\ Lett.$, **A40**, 143.

Hartmann, O. *et al.* (1980). *Phys. Rev. Lett.,* **44**, 337.

Ishida, K. *et al.* (1985). *Phys. Rev. Lett.,* **55**, 2009.

Kadono, R. (1992). In *Meson Science*, ed. T. Yamazaki, K. Nakai, and K. Nagamine, p. 113.
    Amsterdam: North Holland.

Kadono, R. *et al.* (1985). *Phys. Lett.,* **A109**, 61.

Kadono, R. *et al.* (1989). *Phys. Rev.,* **B39**, 23.

Kadono, R. *et al.* (1990). *Phys. Rev. Lett.,* **64**, 665.

Kadono, R. *et al.* (1994). *Phys. Rev. Lett.,* **73**, 2724.

Kadono, R. *et al.* (1999). *Phys. Rev. Lett.,* **83**, 987.

Kagan, Y. (1992). *J. Low Temp. Phys.,* **87**, 525.

Kagan, Yu. and Prokof'ev, N.V. (1990). *Phys. Lett.,* **A150**, 320.

Kagan, Yu. and Prokof'ev, N. V. (1991). *Phys. Lett.,* **A159**, 289.

Karlsson, E.B. *et al.* (1995). *Phys. Rev.,* **B52**, 6417.

Kempton, J.R. *et al.* (1990). *Hyperfine Interactions*, **65**, 811.

Kiefl, R.F. *et al.* (1984). *Phys. Rev. Lett.,* **53**, 90.

Kiefl, R.F. *et al.* (1986). *Phys. Rev.,* **A34**, 681.

Kiefl, R.F. *et al.* (1988). *Phys. Rev. Lett.,* **60**, 224.

Kiefl, R.F. *et al.* (1989). *Phys. Rev. Lett.,* **62**, 792.

Kiefl, R.F. and Estle, T.L. (1990). In *Hydrogen in Semiconductors*, ed. J. Pankove, and N.M. Johnson,
    p. 547. New York: Academic Press.

Kondo, J. (1984). *Physica,* **B125**, 279.

Kondo, J. (1999). *J. Phys. Soc. Jpn*, **68**, 3315.

Kreitzman, S.R. *et al.* (1995). *Phys. Rev.,* **51**, 13117.

Luke, G.M. *et al.* (1991). *Phys. Rev.,* **B43**, 3284.

Meggers, E. *et al.* (1998). *J. Am. Chem. Soc.,* **120**, 12950.

Morozumi, Y. *et al.* (1986). *Phys. Lett.* **A118**, 93.

Nagamine, K. *et al.* (1984). *Phys. Rev. Lett.,* **53**, 1763.

Nagamine, K. *et al.* (2000). *Physica,* **B289–90**, 631.

Nishiyama, K. (1992). In *Meson Science*, ed. T. Yamazaki, K. Nakai, and K. Nagamine, p. 199.
    Amsterdam: North Holland.

Nishiyama, K. *et al.* (1985). *Phys. Lett.,* **111**, 369.

Nishiyama, K. *et al.* (1986). *Hyperfine Interactions,* **32**, 551.

Patterson, B.D. (1988). *Rev. Mod. Phys.,* **60**, 69.

Percival, P. (1980). *Radiochimica Acta*, **26**, 1.

Pratt. F.L. *et al.* (1997). *Phys. Rev. Lett.,* **179**, 2855.

Risch R. and Kehr, K.W. (1992). *Phys. Rev.,* **B46**, 5246.

Roduner, E. *et al.* (1978). *Chem. Phys. Lett.,* **57**, 37.

Roduner, E. *et al.* (1981). *Chem. Phys.,* **54**, 2610.

Roduner, E. (1988). *The Positive Muon as a Probe in Free Radical Chemistry.* Lecture Notes in
    Chemistry 49. Heidelberg: Springer.

Roduner, E. (1999). In *Muon Science*, ed. S.L. Lee, S.H. Kilcoyne, and R. Cywinski, p. 173. Berlin:
    NATO Advanced Study Institute.

Sahoo, N. *et al.* (1983). *Phys. Rev. Lett.,* **50**, 913.

Scheicher, R.H. *et al.* (2001). *Hyperfine Interactions*, **136/137**, 755.

Scheicher, R.H. *et al.* (2003). *Physica B*, **326**, 30.

Shirakawa, H. and Ikeda, S. (1971). *Polym. J.*, **2**, 231.

Srinivas, S. *et al.* (1997). *Hyperfine Interactions*, **105**, 167.
Storchak, V.G. and Prokof'ev, N.V. (1998). *Rev. Mod. Phys.*, **70**, 929.
Su, W.P. *et al.* (1979). *Phys. Rev. Lett.*, **42**, 1698.
Torikai, E. *et al.* (1993). *Hyperfine Interactions*, **79**, 879.
Torikai, E. *et al.* (1994). *Hyperfine Interactions*, **97–8**, 389.
Torikai, E. *et al.* (2001). *Hyperfine Interactions*, **138**, 509.
Van de Walle, C.G. *et al.* (1988). *Phys. Rev. Lett.*, **60**, 2761.
Yamada, K. (1984). *Prog. Theor. Phys.*, **72**, 195.

9

Cosmic-ray muon probe for internal structure of geophysical-scale materials

The spatial profile of the depth or density times length of the substance can be known by observing the manner of penetration of the energetic particle determined by electromagnetic interaction between the incoming particle and the stopping substance, which is called radiography. The most popular radiography is X-ray photography of a human body. As summarized in Table 9.1 and Figure 9.1, among various possible particles, very-high-energy (≥ 100 GeV) muon is the most suitable for measuring the density profile of a large-scale (≥ 0.1 km) substance like a mountain.

As described in Chapter 2, GeV–TeV cosmic-ray muons are constantly irradiating every substance on the earth. Muons arriving vertically from the sky have an intensity of 1 muon/cm^2 per min with a mean energy of a few GeV. The potential use of such high-energy muons to explore the internal structure of large-scale objects has been recognized in the past, with the prime example being the work done by Alvarez (1970), who studied the inside of an Egyptian pyramid in order to find a hidden chamber.

Muons arriving nearly horizontally along the earth's surface with a θ_z slightly less than 90° have a lower intensity on average, but have a higher intensity at energies higher than a few hundreds of GeV, as can be seen in Figure 2.10. For the purpose of probing the internal structures of truly gigantic (geophysical-scale) objects, for example mountains, these horizontal muons are easier to use, provided that the muon flux is reasonably high and that the size of the required detection system can be made realistic. Here, it is essential to eliminate a huge background due to a soft cosmic-ray component caused by shower electrons/positrons and photons which are in part produced during the passage of the cosmic-ray muon through air. The intensity of the soft component is one-quarter of the cosmic-ray muon intensity at $\theta_z \cong 0$, while it is more than 100 times larger at $\theta_z \to 90°$ and typical energy spectrum is from 0.1 to 2 GeV.

9.1 Penetration of cosmic-ray muons through large-scale matter

Here, we consider the conditions required for the efficient use of near-horizontal muons as a probe for the internal structure of gigantic objects. The idea is then extended to incorporate the notions of three-dimensional tomographic measurement of the internal structure and measurements of time-dependent changes occurring within the object. Based on these experimental techniques and other considerations, a new method of predicting volcanic

Table 9.1 Scale of radiography by various particles

Particle	Basic interaction	Penetration characteristics
Electron, X-ray	Electromagnetic	A few meters or less for conversion
Proton, neutron, pion	Strong and electromagnetic	~10 m for absorption
Neutrino	Weak	Earth-size and difficult to detect
Muon	Electromagnetic and weak	100–1000 m and easy to detect

eruptions is proposed by detecting the near-horizontal cosmic-ray muons passing through the active part of a volcano (Nagamine *et al.*, 1995).

The basic idea of this proposed measurement can be explained through the following steps:

1. The cosmic-ray muon energy spectrum and its dependence on vertical angle. Here, we would like to recall the properties of the cosmic ray muons originating in the decay of the pions and the kaons produced through nuclear interactions between primary cosmic-ray protons and the atmospheric air. The energy spectrum of these muons was already discussed in section 2.6 and is summarized in Figure 2.10.

2. Range of cosmic-ray muons through mountain composed of rock. As is also well-known, the energy loss of a charged particle with some energy E(TeV) on passage through matter with a thickness (expressed in terms of density length, that is, density times geometrical length) X (hg/cm^2 = 100 g/cm^2) can be written (Adair and Kasha, 1976) as:

$$dE/dX = [1.888 + 0.077 \ln(E/m_\mu) + 3.9E] \times 10^{-6} \quad \text{(TeV/ g per cm}^2)$$

where the first two terms represent ionization loss and the third term represents stochastic processes due mainly to bremsstrahlung. Choosing E = 1 TeV to yield a relatively small contribution of the logarithmic term, the mean range X can be obtained by integrating the energy-loss formula:

$$X = 2.5 \times 10^3 \ln(1.56E + 1) \text{ (hg/cm}^2)$$

3. Intensity of cosmic-ray muons penetrating through rock with thickness X. Thus, a unique relationship exists between X and the intensity of the penetrating cosmic-ray muons: $N_\mu(E_c(X), \theta_z)$. Once X is given, the minimum energy E_c required for a cosmic-ray muon to penetrate through the thickness X is determined through the X–E relation, and the integrated flux N_μ (E_c, θ_z) is given uniquely. Conversely, for a substance with an unknown thickness X, the measurement of the muon flux N_μ (θ_z) penetrating through the substance with a zenith angle θ_z uniquely determines its thickness in hg/cm^2. The relationship between X and N_μ at different θ_z is summarized in Figure 9.2.

As can be seen from Figure 9.2, small changes in X due to the existence of regions of lower or higher density inside the broadly uniform object lead to differences in N_μ (θ_z); the change in N_μ (θ_z) informs us of the change in X. As a simple example, let us consider a mountain, circular in section with a diameter of 500 m, composed of rock with a uniform

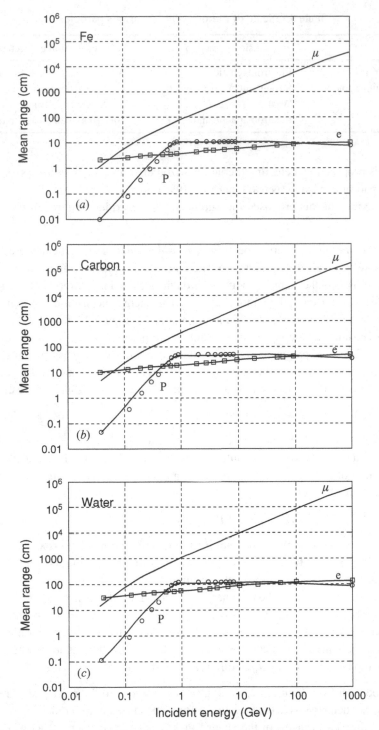

Figure 9.1 Mean range (cm) of protons (p), electrons (e) and muons (μ) through (*a*) iron, (*b*) carbon, and (*c*) water against particle energy (GeV). Here, the mean range is defined as the length of the material where the number of the transmitted particle is attenuated to 50% of the number of the initial injection.

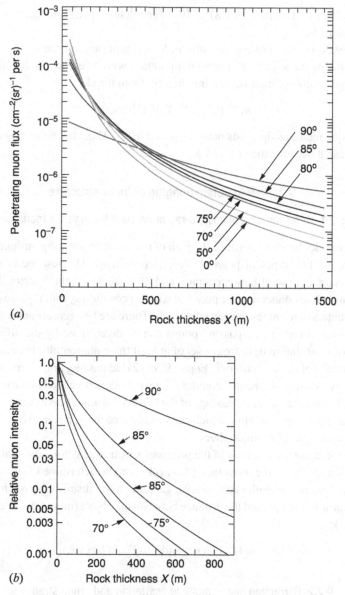

Figure 9.2 (a) Integrated flux of cosmic-ray muons at various zenith angles (θ_z) penetrating through a given thickness (m) of rock with a density of 2.5 g/cm^3. (b) Relative intensity of these muons normalized by the value for zero thickness.

density of 2.5 g/cm^3, and suppose that this mountain contains a cavity or vacancy with a length of 50 m (10% change in X), and that the cosmic-ray muons being observed pass through the mountain with $\theta_z = 90°$. As can be obtained from the $X–E$ formula (see also Figure 3.3) the required energy for a cosmic-ray muon to penetrate through X changes from 0.416 TeV to 0.364 TeV in the vacancy region, yielding a change in N_μ for $\theta_z = 90°$

(Figure 9.2) from 1.61×10^{-6} to 1.87×10^{-6} (cm^{-2}(sr)$^{-1}$ per s); in other words, there is a 16% change in N_μ.

In Figure 9.2, we also present the ratio $n(X, \theta_z)$, representing the relative intensity of cosmic-ray muons transmitted through the mountain with reference to that directly transmitted through "nothing," that is, arriving directly from the sky:

$$n(X, \theta_z) = N(X, \theta_z)/N(0, \theta_z)$$

Again, a unique relationship exists between X and $n(\theta_z)$. Thus, by measuring $n(\theta_z)$ with θ_z known, one can determine the value of X.

9.2 How to obtain imaging of inner structure

9.2.1 Determination of cosmic-ray muon path through the mountain

Now, let us consider how to determine the path of the muon penetrating through the gigantic object. There are two types of practical detection system: (1) telescope of the position-sensitive detector, e.g., scintillation counter; and (2) Cerenkov light detector. In the former case, the straight line connecting the points at which a cosmic-ray muon passes through two (or three) counters determines the particle's path. There are two representative examples of position determination: (1) the passing points can be determined by the difference in the arrival time of scintillation light from a set of at least three photomultipliers attached to the edges (or corners) of each scintillator (Figure 9.3); (2) the passing points can be determined more straightly by using segmented counter arrays in both vertical and horizontal directions (Figure 9.3). In the latter case, the passage of the relativistic muon leaves a track of Cerenkov radiation in a selected gas or liquid which can be collected using a concave reflective mirror focused on an array of photomultipliers.

In the case of counter telescopes of the position-sensitive detectors, the spatial resolution of the detection system at the mountain, $(\Delta X, \Delta Y)$, can be determined from the resolution of the intersection points with each counter, $(\Delta x, \Delta y)$, the distance between the counters comprising the telescope, ℓ, and the distance between the object (mountain) and the detector, L (Figure 9.3):

$$\Delta X = (L/\ell)\Delta x \quad \text{and} \quad \Delta Y = (L/\ell)\,\Delta y$$

9.2.2 Correction due to multiple scattering and range straggling

For the energy region where ionization is the dominant energy-loss process, the effect of multiple scattering experienced by the muons during their passage through the mountain can be estimated by Molière's theory. In fact, this is the case for muons penetrating through rock with energies well below 1 TeV. The effect is more serious for muons in the low-energy region where the energy loss corresponds to a large fraction of the particle's original energy. According to the Monte Carlo calculation, it was found that, given the energy spectrum of cosmic-ray muons for θ_z above 70° (i.e., between 70° and 90°), the overall angular spread can be maintained below 5 mr for muons passing through a mountain 500 m thick.

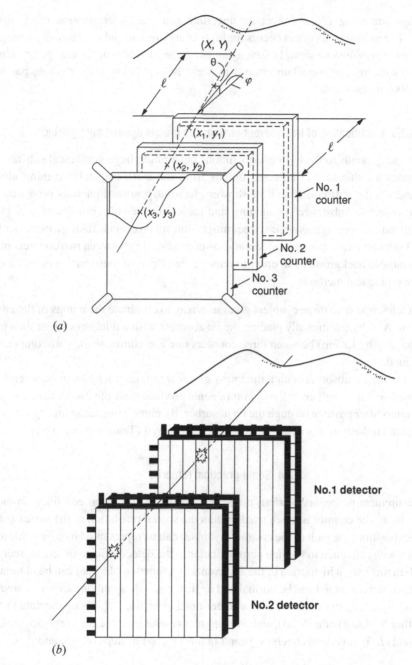

(a)

(b)

Figure 9.3 (a) Counter telescope comprising three plastic scintillators used for the Mt Tsukuba measurement. (b) The detection system comprising two segmented plastic counters used for the Mt Asama measurement, where iron plates, as described in the text, are to be placed in between the two counters in order to eliminate the soft cosmic-ray component by multiplicity cut.

The range-straggling effect causes an uncertainty in the determination of X from N_μ (θ_z, ϕ). The most serious effect occurs for high-energy muons, where bremsstrahlung is the dominant energy-loss process. In fact, again after consideration of the energy spectrum of cosmic-ray muons, the overall uncertainty in X can be kept below 1% for muons passing through a 500-m mountain.

9.2.3 Identification of the relevant cosmic-ray muons against backgrounds

The major backgrounds to the high-energy muons penetrating large geological substances like volcanoes are due to: (1) reflected muons from the earth originally coming along $\theta_z \approx 0°$; and (2) the soft component of shower electrons/positrons/photons produced in the air, the objective substance (mountain), and the earth; the soft component may give substantially misleading signals against true straight-line muon events. Taking a two-counter telescope, both muon reflection and electron shower produced in between two counters may cause unavoidable backgrounds. In order to remove the effects of these backgrounds, there might be two practical methods:

1. Using a telescope with three counters gives us a way to eliminate the muons of the earth reflection. Also by additionally placing the Fe absorber with a thickness larger than one radiation length (1.8 cm) between three counters one can eliminate the soft-component background.
2. By placing the Fe absorber with a thickness of a few radiation lengths in between two segmented counters, soft cosmic-ray components produce multiple events in one of the two counters after passage through the Fe absorber. By eliminating these multiple events a substantial reduction in soft component can be achieved (Tanaka *et al.*, 2001).

9.2.4 Some practical remarks

In the measurement procedure, various muon paths $X(\theta_z, \phi)$ are recorded. Since, in most cases, the size of the counter is much smaller than the spatial resolution of the vertex point at the object position, the path of the cosmic-ray muon can be represented by azimuthal and polar angles with reference to the line perpendicular to the detector plane (θ_z, ϕ), as seen in Figure 9.3. In this way, a histogram of the N_μ events as a function of (θ_z, ϕ) can be obtained. The N_μ (θ_z, ϕ) thus obtained can be normalized against cosmic-ray muons arriving directly from the sky by taking the ratio $n(\theta_z, \phi)$, as mentioned in section 9.1. Using the data in the form of either N_μ (θ_z, ϕ) or n (θ_z, ϕ), and with angular resolution $\Delta\theta_z$, $\Delta\phi$ corresponding to $(L/l)\,\Delta y$ and $(L/l)\,\Delta x$, respectively, we can obtain $X(\theta_z, \phi)$ in steps of $\Delta\theta_z$ and $\Delta\phi$.

9.2.5 Tomographic imaging

The imaging of inner structure by cosmic-ray muon should be referred to as a Röntgen photograph of the gigantic substance by replacing X-ray with cosmic-ray muon. The main technical difference is due to adoption of an angular-dependent counting system instead of the parallel beam usually employed in X-ray imaging. By using more than two counters placed at different angular geometry, one can obtain three-dimensional tomographic imaging of the inner structure.

9.3 Example of counter system and data analysis

In order to realize cosmic-ray imaging of a large-scale substance like a volcanic mountain, several counter systems with position-sensitive and multiple natures have been tested. Among them, two representative examples are described below.

9.3.1 Analog three-counter system

In order to identify the path of cosmic-ray muons, a threefold telescope of plastic scintillation counters was employed in the earliest test experiment (Nagamine et al., 1995). As shown in Figure 9.3, the path of a muon can be determined by connecting the three intersection points ($x1$, $y1$; $x2$, $y2$; $x3$, $y3$). The use of a threefold telescope makes it easier to eliminate the background events caused by low-energy muons with an incident θ_z of 0° which are reflected from the earth between the counters. In order to maximize the number of particles counted, the first set of measurements was carried out in compact geometry with $\ell = 1.5$ m distance between no. 1 and no. 3 counters.

Each plastic scintillator has a shape of 127×127 cm area and 3.0 cm thickness. At each of the four corners, a photomultiplier is placed with a minimum volume of Lucite interconnector between the scintillator and the photomultiplier. The point where the muon hits the counter is determined by detecting the arrival time of scintillation light at each photomultiplier and noting that the delay from the initial muon impact is 5.3 ns times the distance (m) between the impact point and the photomultiplier position. The change in light yield seen in the pulse height from the photomultiplier can also be used as a complementary source of information on the impact point. In most measurements, a spatial resolution of ±2.5 cm was obtained in the determination of the muon impact position.

The time-to-digital converter (TDC) circuit determines all the timings from the 4×3 photomulipliers with reference to a starting pulse which is actually taken from the mean of the pulse times from the four photomulipliers of the first counter. Data-taking is initiated by the event-trigger pulse, which is a coincidence signal from all of the 4×3 pulses.

9.3.2 Segmented two-counter system

As shown in Figure 9.3, the other detection system is composed of two segmented detectors. Each segmented detector consists of x and y planes. Each plane consists of an array of 10 counters. Each counter is composed of a plastic scintillator 10 cm wide $\times$ 100 cm long $\times$ 3 cm thick and a photomultiplier tube (PMT). Ten counters were arranged in both the x and y planes, and a spatial resolution of ±5 cm was realized to determine the muon hitting position. The two segmented detectors were placed at a distance of 1.5 m to achieve ±66 mrad angular resolution. In order to remove a contribution from electrons in the soft cosmic-ray component, two 5-cm-thick iron plates were placed between the two counters and multiple events produced at the iron plates and detected in the second counter were rejected.

High voltage applied to each PMT of each counter was adjusted so that the deviation of the event rate was less than $\pm 10\%$ from the average. TDCs measure all the timings of the 40 PMTs. The start signal and the event trigger to TDC were generated by a coincidence of signals of any PMT of each of the three different planes in a certain time gate. The electronic noises such as dark current were almost entirely rejected by requiring the signal coincidence of each layer. The data from each TDC were transferred to the computer controlled by the data acquisition system.

9.3.3 Data-taking and analysis

All of the data-taking and online monitoring is carried out using a standalone workstation (initially a VAX 3200, later an IBM PC computer) with the data-acquisition system of EXP95 (Nakamura et al., 1997).

As for the three-counter system, the angles (θ_z, ϕ) of the muon path are obtained from $(x1, y1)$ and $(x3, y3)$, with $(x2, y2)$ used as extra data to test the straightness of the trajectory by applying a linear fit to the three points, and the relative timings from the three scintillators are used to identify the directional sign of the muon path, providing $N_\mu(\theta_z, \phi)$ for FWD (forward: corresponding to muon paths from counter no. 1 to counter no. 3), $N_\mu(\theta_z, \phi)$ for BWD (backward: corresponding to those from counter no. 3 to counter no. 1), and $N_\mu(\theta_z, \phi)$ for BG (background: corresponding to random background without timing appropriate to any trajectory through the counters).

As for the two-counter system, the data from the 40 TDCs were converted into histograms on the online monitor as follows:

1. A cosmic-ray muon makes an event trigger and, thus, a real muon signal is observed at ·a certain time in the TDC time spectrum.
2. The spatial position of a cosmic-ray muon hitting each segmented detector was determined from a combination of the signal in the x and y plane.
3. From the straight line connecting the positions on the two segmented detectors, the arriving angles were determined.
4. When more than two signals from the same plane coincided in the time gate, such an event was discarded as the signals due to a soft component (multiplicity cut).
5. The relative timing from two detectors was used to identify the muon path for FWD and for BWD as well as for BG.

9.4 Results of some feasibility studies

In order to confirm the feasibility of the presently proposed method, test experiments employing a simple set-up to detect near-horizontal cosmic-ray muons have been conducted for Mt Tsukuba and Mt Asama. The goals of these test experiments were as follows: (1) to check whether N_μ does indeed depend on X, and to establish the possibility of X determination from N_μ; and (2) to see structure in the inner volcanic crater from the outside.

9.4.1 Mt Tsukuba experiment

In the first feasibility test experiment, one three-counter system of the analog type was placed at the foot of Mt Tsukuba, at an elevation of 150 m from sea level and at a distance of 2.0 km from the midpoint of Mt Tsukuba's two peaks (Otokoyama 870 m and Onnayama 876 m).

The results of the Mt Tsukuba measurements over 33 days are shown in Figure 9.4 in the form of a two-dimensional histogram of $N_\mu(\theta_z, \phi)$, corrected for angular acceptance and with the result presented at various levels of discrimination with respect to low-rate events. The result demonstrates the following important features regarding the structure of Mt Tsukuba as explored by cosmic-ray muons. By selecting the low-event-rate cut level to be around 20% of the maximum rate, the outer profile of the mountain can clearly be seen, while there still exists a significant fraction of cosmic-ray muons which penetrate the mountain, representing a probe for its inner structure. As is evident in Figure 9.4, the following features of the internal structure of Mt Tsukuba were observed: (1) the overall densities in both the peak region and the lower-altitude part of the mountain are close to $2.0 \, \text{g/cm}^3$; and (2) there seems to be a less dense part in the region of the interpeak midpoint.

9.4.2 Mt Asama experiment

Two two-counter systems of the segmented type were placed at the foot of Mt Asama located in Gunma prefecture, Japan; its elevation is 1400 m above sea level and its horizontal distance is 2750 m from the center of the crater at the peak of Mt Asama (2570 m in elevation), where the crater is 300 m in diameter and 228 m in depth. A Monte Carlo simulation was performed for Mt Asama with a given density distribution for comparison with the measured data over 90 days. Based on continuous variations of a mountain shape, the method of spline interpolation was taken in order to obtain more precise spatial segmentation. The histograms shown in Figure 9.5 are the Monte Carlo simulations of several crater conditions and the data obtained where all the figures are presented after subtraction of filled-crater Monte Carlo simulation. The event increase was seen at the angle corresponding to the position of the crater. The absolute value of obtained data agrees well with the Monte Carlo simulation where volume occupancy by magma in the crater is less than 30%.

9.5 Prospects for volcanic eruption prediction

Among various conceivable future applications of this method of probing the internal structure of mountains or other geophysical-scale objects, we propose here an application in the prediction of volcanic eruptions. For this purpose, let us consider how we can relate the process of volcanic eruption to the interior structure of the mountain. As shown in Figure 9.6, a

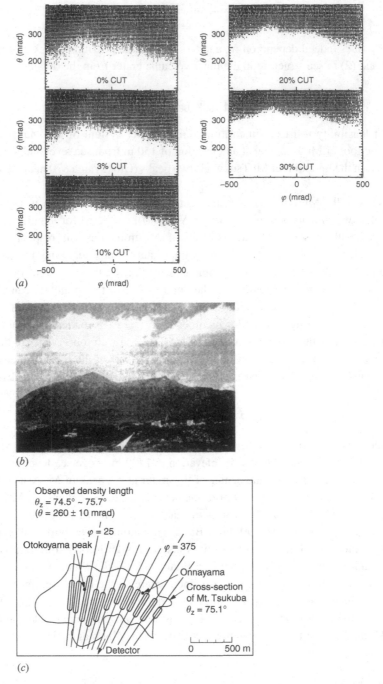

Figure 9.4 (a) $N_\mu(\theta, \phi)(\theta = 90° - \theta_z, \theta_z$: zenith angle) histogram obtained in the Mt Tsukuba measurement with varying degrees of discrimination of the lower event rates: no discrimination, 3% discrimination, 10% discrimination, 20% discrimination, and 30% discrimination, where the maximum event point in the histogram is taken as 100%. The data are for a measurement with the 1.5 m distance counter facing the point central between the two peaks. (b) A photograph of Mt Tsukuba along the detector direction is shown, with the counter location indicated by an arrow. (c) Cut-view presentation of typical data for $X(\theta_z, \phi)$ with $\theta_z = 74.5°$–$75.7°$ and $\theta = 260 \pm 10$ mrad; a uniform density of 2.0 g/cm^3 is assumed. A cut-view of the mountain along θ_z obtained from the map is also shown. The position of X along the line is arbitrary.

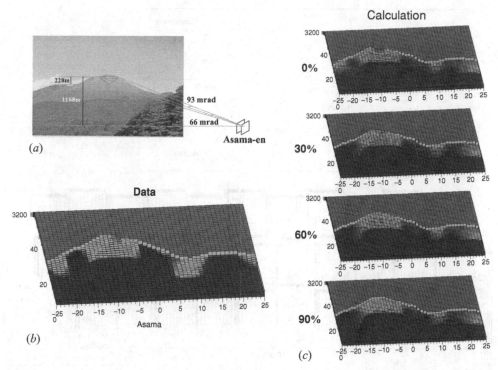

Figure 9.5 (*a*) Conceptual three-dimensional view of Mt Asama and (*b*) experimental data emphasizing the effect of a crater in comparison with (*c*) Monte Carlo simulations for different levels of filling of a substance like magma in the crater. These figures are the subtraction of the calculation for completely filled crater from those for various-level conditions using spline interpolation. The increase in number at the top region indicates the existence of a crater.

volcanic eruption is likely to be preceded by a change in density along the crater or magma channel inside the upper part of the volcano. In order to simplify the situation, let us consider an extreme situation in which the volcanic eruption process involves magma with a density similar to volcanic rock (density $= 2.5$ g/cm^3) passing through an initially vacant channel (density $= 0$ g/cm^3).

Based on test measurements, one can now quantitatively estimate how this cavity model of volcanic eruption can be observed by means of cosmic-ray muons. We consider a realistic enlargement of the original counter system. The following factors are important in the design of the full-scale system: (1) efficiency and speed of data-taking; (2) cost required for all the equipment including detectors, data-taking electronics and computer, and for setting up all the equipment; (3) availability of space to accommodate the whole set-up. Taking all these factors into consideration, we propose here a detector system with a sensitive area of 20 m^2. As is seen in Figure 9.6, the proposed system is scaled up in the area from the system used in the Mt Asama measurement by a factor of 20. The detector will cover an area which is two times taller (2.0 m) and five times wider (10 m) than the one used in the Mt Asama experiment.

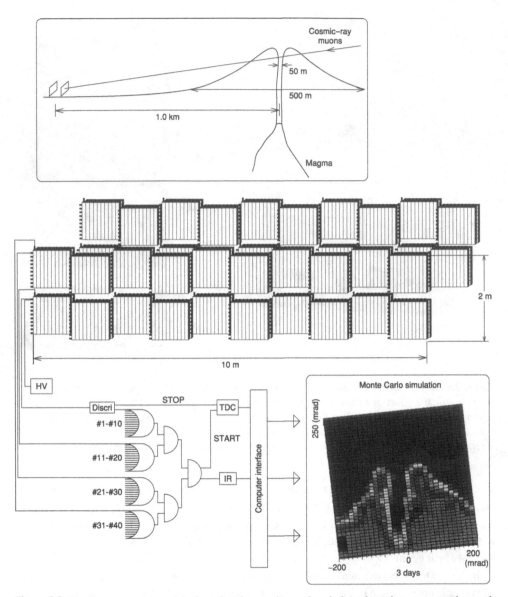

Figure 9.6 Cavity movement model of a volcanic eruption and scaled-up detection system to be used to predict volcanic eruptions. HV, high-voltage power supply; TDC, time-to-digital converter; IR, interrupt register.

To simulate the results of measurements using such a system, a Monte Carlo calculation was performed for a mountain with realistic shape, size, and density parameters. If a cavity like a crater of 50 m diameter exists in the upper region of the volcano with a thickness of 500 m, the result of the Monte Carlo calculation demonstrates that probing the cavity from an observation point 1.0 km from the volcanic eruption channel can be carried out within

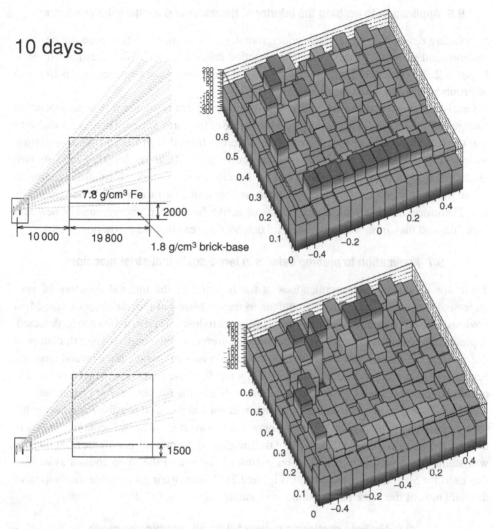

Figure 9.7 Simulation calculation of transmitted cosmic-ray muon intensity through a blast furnace with different thickness (2 m and 1.5 m) of brick-base, where the detection system used in the Mt Asama experiment is employed in the geometry (10 m from the furnace wall) shown at the left-hand side.

a reasonably short time (a few days). Thus, by using an enlarged detector with an area of 20 m² located at a distance of 1 km, facing the mountain with a zenith angle of 80°, and with a spatial resolution of 10 m × 10 m at the mountain, an anomaly of 10% in X can be detected in a few days.

Thus, if a change in X of this magnitude associated with volcanic eruption occurs over a time scale of a few days, the eruption can be predicted by time-dependent measurement of the transmitted cosmic-ray muon intensity through the region around the crater or magma channel.

9.6 Application to probing the interior of the earth and earthquake prediction

By placing detector systems deep underground, one can measure the three-dimensional underground structure of the earth. The detection efficiency can again be estimated using Figure 9.2. The practical limit of the thickness of rock which can be penetrated is likely to be around a few km, perhaps 10 km at most.

One of the important applications of this technique might be to measure time-dependent changes in active faults in the underground structure of the earth's crust. There is a possibility that the overall density of the active fault is different from that of the surrounding normal region – close to a 20% smaller density at the active fault. Thus, the location and structure of the active fault deep underground can be detected. So far, most information relating to active faults only comes from measurements at or near the surface (within a few meters or so). Therefore, information on movements of active faults deep underground is new and valuable, and may in fact also be a helpful database for earthquake prediction.

9.7 Application to probing defects in large-scale industrial machinery

There are several possible applications of the imaging of the internal structure of very large-scale objects with cosmic-ray muons to the problems in industrial apparatus. Most obviously, internal defects or failures in a large-scale industrial construction can be detected. A practical and useful example might be the measurement of the bottom-layer thickness of the massive blast furnace used in the iron-production industry where the iron and coke are heated by hot air to produce molten iron. Thus far, the base part constructed from bricks cannot be monitored, and so the end of the useful life of the furnace cannot be estimated in advance. Once an optimized detection system is chosen and placed sensibly, monitoring the furnace base using the present technique may serve to avert tremendous economic losses in the industrial world. A typical example of the imaging of the bottom-part of the blast furnace was studied by a simulation calculation, assuming the use of the same counter system as that used for Mt Asama measurement (Figure 9.7), demonstrating a possible monitoring of the thickness of the brick base within a reasonably short time (10 days).

9.8 Multiple scattering radiography with cosmic-ray muon

By tracking the scattering angles of individual incoming and outgoing muons using a set of two high-resolution position-sensitive detectors like drift chambers, one can develop a new type of radiography for the objective placed in between two sets of detectors. Such radiographic imaging has recently been developed (Borozdin et al., 2003).

REFERENCES

Adair, R.K. and Kasha, H. (1976). In *Muon Physics 1*, ed. V.W. Hughes, and C.S. Wu, p. 323. New York: Academic Press.
Alvarez, L.W. (1970). *Science, 167*, 832.
Borozdin, K.N. *et al.* (2003). *Nature, 422*, 277.
Nagamine, K. *et al.* (1995). *Nucl. Instr., A356*, 585.
Nakamura, S.N. *et al.* (1997). *Nucl. Instr., A388*, 220.
Tanaka, H. *et al.* (2001). *Hyperfine Interactions, 138*, 521.

10

Future trends in muon science

Scientific research with muons began with the discovery of muons in cosmic rays in 1941. Therefore, it is definitely young, perhaps even in its infancy. Some possibilities for the future have already been given at the end of some chapters (e.g., Chapters 5, 8, and 9). During the twenty-first century, assisted by the realization of new intense hadron accelerators and by a substantially improved muon beam production method, muon science will make further new and exciting strides in its development. In this chapter, several possible future scientific programs will be presented simply according to the author's personal view.

10.1 Nonlinear muon effects

At present, most muon science experiments have been conducted in the dilute limit of muon intensity, where a given injected muon is not interacting with any other muon. There, the rate of all the muon-associated signals $\dot{N}_{\text{sig.}}^{(\mu)}$ is proportional to the muon intensity (in a unit time interval and a unit volume); $\dot{N}_{\text{sig.}}^{(\mu)} \propto \dot{N}_{\mu}$. This situation can be seen more clearly in the form of instantaneous spatial density of muons stopping in matter, as shown in Figure 10.1. Obviously, pulsed muons can provide substantially higher spatial density.

It is interesting to foresee what kind of new physics can be disclosed by the increasing intensity of pulsed muons. There, one can expect that one muon may interact with other muons. Thus, the nonlinear term becomes significant: $\dot{N}_{\text{sig.}}^{(\mu)} \propto \dot{N}_{\mu} + \dot{N}_{\mu}^2 + \cdots$.

It is important to estimate how much correlated events become significant under how much muon intensity and under what kind of correlation among muons. One way to estimate this is to consider the interaction cross-section σ_c of one incoming muon flux $j(N_1)$ with the other stationary muons with the density of $\dot{N}_2$. They encounter each other with a repetition frequency of $f_n(\text{s}^{-1})$. The rate of nonlinear muon-correlated events $\dot{N}_c^{(\mu)}$ becomes as follows: $\dot{N}_c^{(\mu)} = [j_1(N_1) \cdot \tau_{\mu} \cdot \dot{N}_2 \cdot \sigma] \cdot f_n$. For example, for $j(N_1)$ of 1 μA (0.6×10^7/s at a duty factor of 10^{-6}) in 2 mm diameter ($2 \times 10^8 \text{cm}^{-2}$/s) with N_2 of 10^8 in the interaction volume and with a cross-section σ of 10^{-16} cm^2 corresponding to the range of interaction of the order of 1 Å and with f_n of 20 s^{-1}, one can expect $\dot{N}_c^{(\mu)}$ of 0.8×10^{-4}/s or one event per 3 h.

The other way of estimation is to consider a rate of accidental coincidence of the intense muons of event rates $\dot{N}_1$, $\dot{N}_2$ within a correlation time τ_c: $\dot{N}_c^{(\mu)} = 2\tau_c \dot{N}_1 \cdot \dot{N}_2$.

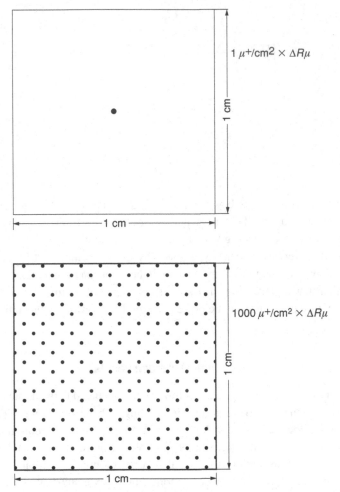

Figure 10.1 Instantaneous spatial density of muons stopping inside the matter in the case of Paul Scherrer Institute (PSI) (*a*) and High Energy Accelerator Research Organization–Meson Science Laboratory (KEK-MSL) (*b*). In the case of Institute of Physical and Chemical Research–Rutherford Appleton Laboratory branch (RIKEN-RAL), the density of muon stopping becomes 20 times larger still.

An important application of these nonlinear effects exists in various fields of muon science. Some representative examples are as follows:

1. Paired μ^+/Mu diffusion, which may be helpful to explore the origin of paired hydrogen diffusion, which is a well-known phenomenon in hydrogen diffusion in metals at low temperature.
2. The nonlinear phenomena in muon catalyzed fusion; some descriptions are given in section 5.6.
3. Formation of the $\mu^+\mu^-$ atom; details are given in section 10.3.

10.2 Production of muonic antihydrogen and CPT theorem

Including antiproton $\bar{p}$, there are four types of hydrogen atoms allowing species involving μ^+ and μ^-: these are the conventional H atom (e^-p), the corresponding anti-atom $e^+\bar{p}$ (known as antihydrogen, $\bar{H}$), and the two muonic counterparts, μ^-p and $\mu^+\bar{p}$. If a method of generating antihydrogen $\bar{H}$ ($e^+\bar{p}$) is established, it is widely discussed that a high-precision spectroscopic measurement on $\bar{H}$, in comparison with the corresponding results for H, may contribute to the verification or falsification of the CPT conservation law (where CPT refers to the product of the three symmetry operations of charge transformation, parity inversion, and time reversal).

The advantage of the use of the μ^-p, $\mu^+\bar{p}$ pair is obvious. If the CPT-violating interaction is short-range (with an extremely massive exchange boson), such an effect can be seen more easily in the (μ^-p, $\mu^+\bar{p}$) in comparison with that in the (e^-p, $e^+\bar{p}$) case since the atomic size becomes smaller by $1/207$.

Since intense slow μ^+ and Mu beams will soon become available, it will be possible to produce $\mu^+\bar{p}$ through, e.g., the following reaction: $Mu + \bar{p} \rightarrow \mu^+\bar{p} + e^-$, i.e. thermal Mu and $\bar{p}$ reaction with the energy of $\bar{p}$ optimized for the binding energy of the final state $\mu^+\bar{p}$ ($E_{q.s.}(\mu^-\bar{p}) \sim 2.8$ keV).

Keeping this aim in mind, it is rather a shame for the present muon physicists that there has been no successful high-precision measurement such as a laser resonance experiment on muonic hydrogen (μ^-p). Apart from the comparison to ($\mu^+\bar{p}$) mentioned above, the system presents many fundamental interests, such as the questions of proton polarizability and vacuum polarization.

As for the energy levels of (μ^-p) shown in Figure 10.2, there have been several proposals for laser resonance spectroscopy. One distinguished example is measurement of the 2S Lamb shift ($2^5P_{3/2} - 2^3S_{1/2}$) of muonic hydrogen proposed at the Paul Scherrer Institute (PSI) (Kottmann et al., 2001). There, a laser resonance spectroscopy experiment is planned to a precision of 30 p.p.m. by employing a $\lambda \sim 6$ μm laser produced by multiple Raman process of 708 nm laser. Another example is (μ^-p) (3d $-$ 3p) with $\Delta E = 0.006$ eV and $\lambda = 188$ μm (Hauser, 1996). For this purpose, a variable-frequency free electron laser is required, together with intense μ^- stopping in low pressure (below 20 mbar) hydrogen gas. Since 88% of the splitting is due to the vacuum polarization, one can determine the vacuum polarization with an accuracy of 100 p.p.m., which can be compared to the presently available value with an accuracy of 0.1% from either $(g - 2)_\mu$ or $(g - 2)_e$.

Another example is the (μ^-p) ($n = 1$, hfs) with $\Delta E = 0.183$ eV and $\lambda = 6.8$ μm. As described by K. Kato (private communication), an appropriate laser source with frequency variability might be feasible. The major difficulty in this measurement is how to detect the resonance signal. A polarized (μ^-p) ($n = 1$) state can in principle be produced by spin-exchange collision between a spin-polarized Kr atom and unpolarized (μ^-p), as commonly used to obtain polarized radioactive nuclei, and muonium. The conditions, i.e., the Kr concentration in H_2 gas and the total pressure of the gas mixture, must be optimized so that μ^- transfer from p to Kr is minimized and spin polarization transfer is maximized. Another possibility might be a brute-force low-temperature nuclear polarization of ortho H_2 (80%

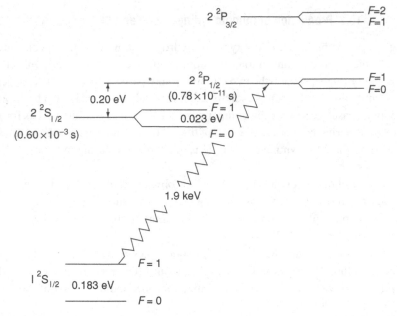

Figure 10.2 Energy levels of μ^-p or $\mu^+\bar{p}$ under CPT (charge transformation, parity inversion, and time reversal) conservation. Life times of 2p and 2s states are given by Hughes and Kinoshita (1977)

polarization under 15 T at 0.015 K). After generating the spin repolarized (μ^-p) in its $F = 1$ state, the hfs resonance can easily be detected by observing the destruction of the decay electron asymmetry with reference to the spin polarization axis.

10.3 The $\mu^+\mu^-$ atom

The muon can participate in the formation of a variety of atomic states; these include states previously discussed, such as (μ^+e^-) (and its corresponding antiparticle pair (μ^-e^+)), (μ^-Z), and more obscure states such as ($\mu^+\pi^-$). Among all of these, the ($\mu^+\mu^-$) atom is the most difficult to produce. The physical properties of ($\mu^+\mu^-$) atoms can be estimated using knowledge related to positronium (e^+e^-), muonium (μ^+e^-), and other systems similar to it. The ($\mu^+\mu^-$) atom is an atomic state formed by a pair of heavy "structureless and point-like" leptons. It can thus in principle serve as a test of quantum electrodynamics without size corrections, and furthermore, because of the large masses of both constituents, important effects such as the weak interaction correction are highly enhanced in this system. Using the formalism developed for muonium and light μ^- atoms together with the knowledge of the masses of the weak bosons ($M_w = 80.8$ GeV, $M_z = 92.9$ GeV), typical weak corrections for this system can be found to be substantially larger than those expected for presently available species such as muonium; the weak correction to the muonium E_{hfs} is only 0.02 p.p.m.

Previously, as a possible avenue of formation of ($\mu^+\mu^-$), the following reaction of (μ^-p) + $\mu^+ \rightarrow$ ($\mu^+\mu^-$) + p has been considered theoretically (Ma *et al.*, 1985). In this case, the maximun cross-section of the order of 10^{-20} cm^2 is predicted at about 2.2 keV μ^+ energy. Now, an interesting collision experiment can be envisaged between a slow μ^- beam and a thermal Mu beam (see section 2.2), both to be produced with pulsed time structure. Then one can expect to produce the ($\mu^+\mu^-$) atom through the following reaction with a larger cross-section: $\mu^- + (\mu^+e^-) \rightarrow (\mu^+\mu^-) + e^-$. Detailed consideration of this approach is now in progress.

10.4 Muonium free drop and lepton gravitational constant

Gravitational force and gravitational mass have been studied with considerable effort all over the world ever since Newton's famous observation of the falling apple. The apple, as well as most of the other falling bodies around us, consists of atoms whose weight is due to the mass of the nuclei, composites of quarks. Thus, all the existing knowledge relating to gravitational mass is so far restricted to the masses of many-body systems of quarks. No knowledge exists in relation to the gravitational mass of leptons such as e, μ, or τ. Of course, the inertial mass of leptons is known with a high accuracy, e.g., up to 1 p.p.m. for the mass of μ^+ (see Chapter 1). Therefore, once the gravitational masses of the leptons are known, the difference between the two types of lepton mass, if any such difference exists, will be known for the first time.

Among the candidates for lepton free-drop experiments – which include the electron (e$^-$) droplet, positronium (e$^+$e$^-$), and other systems – we believe that muonium will be the appropriate candidate, because of less probability of annihilation and the charge neutrality.

Since μ^+ in vacuum has a lifetime of 2.2 µs, the free-drop distance of muonium $L_{FD} = (1/2)\,gt^2$ (g = gravitational constant) is small: for $t = 50$ µs (22.5 τ_μ, μ^+survival of 1.7 × 10^{-10}) $L_{FD} = 0.012$ µm (120 Å); for $t = 22$ µs (10τ_μ, μ^+, survival of 4 × 10^{-5}), $L_{FD} = 0.0024$ µm (24 Å). Use of an extremely intense pulsed μ^+ beam at least on the order of 10^{10} muons/pulse results in survival of $10^5\mu^+$ at 10 τ_μ and survival of 20 μ^+ at 20 τ_μ.

On the other hand, muonium at a temperature T (K) will undergo thermal motion with some characteristic distance: $L_{th} = t\sqrt{2E_{th}/m_\mu} = 0.41 \times 10^5 \times t\sqrt{T}$ cm. For $t = 22$ µs, $L_{th} = 0.9 \times \sqrt{T}$ cm. Therefore, in order to obtain the condition $L_{th} \approx L_{FD}$, we need to cool muonium down to a temperature of at least 10^{-6} K (1 µK). Otherwise, the free-drop behavior is masked by the thermal motion of Mu.

As described in Chapter 2, thermal muonium is generated either through a thermionic process from the surface of a suitable hot noble metal or through a thermal collision process in SiO$_2$ powder. The efficiency of thermal Mu production can be estimated as a ratio of the thermal diffusion length ($\sqrt{DT}$) and the range–width of either μ^+ in metal or Mu in SiO$_2$ (ΔR_μ), both inside the stopping material for thermal Mu generation. Using an intense slow μ^+ beam in the energy range of some keV with $\Delta R_\mu = 50$ Å, one can obtain a complete conversion of μ^+ to thermal Mu by reflecting the slow μ^+ from the surface of either a noble metal or SiO$_2$ layer cooled to liquid He temperature. The velocity distribution of the thermal Mu produced is expected to correspond to a temperature of 4.2 K

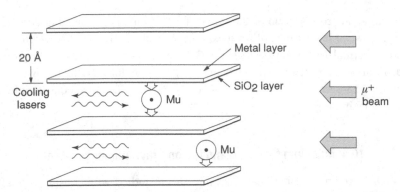

Figure 10.3 Possible arrangement for the muonium free-drop experiment employing a multilayer arrangement of SiO_2/metal sheets.

with the r.m.s. velocity of 1 mm/μs. The lasers for the three-dimensional cooling of Mu could then be applied at a few millimeters above the surface in the geometry shown in Figure 10.3. Although there is some evidence that the thermal Mu from the SiO_2 powder might be chemisorbed to the SiO_2 surface, the strong vacuum ultraviolet (VUV) laser described below will easily desorb Mu from the surface through a laser aberration process. It is well known that the kinetic energy of atoms can be efficiently reduced by the method of Doppler cooling using laser radiation tuned to the lower-frequency-half of the Doppler broadened absorption line. Using three pairs of counterpropagating beams in the optical molasses configuration, three-dimensional cooling can be achieved, and this method can be applied for cooling down to 11 mK of the precooled Mu atoms at 4.2 K (P. Bakule, private communication). The low temperature required such as 1 μK may possibly be reached by using additional techniques such as r.f.-assisted evaporation cooling in a magnetic trap, which was recently used to cool atomic hydrogen (Fried *et al.*, 1998).

An artificial multilayer consisting of a sheet with a metallic surface on one side and a SiO_2 surface on the other can be made with an intersheet distance of, say, 20 Å. Mu formed at the SiO_2 side of one sheet can drop freely to the metallic side of the next only if the SiO_2 side faces down and only if the conventional gravitational force acts on muonium similarly to normal matter (Figure 10.3). By adjusting the intersheet distance and the multilayer orientation one can measure the value and sign of g for muonium.

10.5 Advanced neutrino sources with slow μ^+

There is a strong request from the particle physics community to construct an intense and high-quality neutrino source mainly for the purpose of neutrino mass determination. The present idea of a realistic scheme of a neutrino factory comprises intense proton accelerator and high-acceptance pion collector with a sufficient decay section for $\pi\mu$ conversion, followed by a muon cooling, muon acceleration, and a muon storage ring.

One of the most important applications of the intense ultraslow μ^+ sources described in Chapter 2 would be for advanced sources of muon neutrinos ($\bar{\nu}_\mu$; Nagamine, 1999). Starting with a primary source of high-intensity, low-emittance ultraslow μ^+, the particles would be promptly accelerated up to more than 100 MeV; at these energies, special relativity lengthens the particle lifetime considerably, and the loss rate of muons during the subsequent acceleration process remains small. With the installation of an appropriate decay section for the accelerated μ^+ like a racetrack type storage ring, intense high-quality beams of muon and electron neutrinos are produced via $\mu^+ \rightarrow \bar{\nu}_\mu + \nu_e + e^+$. All of these proposed scenarios are summarized in Figure 10.4.

The important key factors for the realization of such muon acceleration-based advanced neutrino sources can be summarized as follows:

1. Quality of accelerated muons. With ultraslow μ^+ generated by laser resonant ionization of thermal Mu from a hot metal surface placed after the super-super muon channel, one can expect a μ^+ source with intensity greater than 10^{10} s^{-1}, with an extremely small phase space (0.2 eV $\times$ (cm)^2). The capture efficiency of the adjacent linear accelerator and the loss due to μ^+ decay then become the limiting factors.
2. Muon decay section. Since the decay length of the accelerated muons is fairly long (L_μ(m) $\approx 4.7\, p_\mu$ (MeV/c)), we require some quite ingenious designs for the muon decay section, with long distance and good confinement. One attractive idea is a racetrack-type muon storage ring.
3. Quality of the neutrino beam produced. The neutrino beam produced via decay of the accelerated muons has properties which are determined by the kinematics of the three-body muon decay. The energy spread and the dimensions of the decay cone are the limiting factors.
4. Neutrino beam monitor. The arrival timing of the neutrino and the quality of the neutrino beam can be monitored via a charged particle appearance reaction such as $\bar{\nu}_\mu \rightarrow \mu^+$. The cross-section for this type of reaction becomes larger at higher energy, making detection easier for high-energy neutrinos.

All of these properties for the case of the proposed system optimized for a 0.8 GeV $\times$ 300 μÅ proton synchrotron (Nagamine, 1999) become $1.0 \times 10^{10} \mu^+$ with the slow muon generator and $0.5 \times 10^{10} \bar{\nu}_\mu$ via full conversion in the racetrack storage ring.

There are several important applications for such an advanced neutrino beam. Some distinguished examples are given here:

1. Neutrino oscillation, where, for the appearance of ν_τ to be detected efficiently, the acceleration of the muon up to 6 GeV is needed.
2. Application of neutrinos to geophysics, where, by employing the advanced neutrino beam proposed here, time-dependent changes in the earth's crust structure, e.g., the movement of an active fault deep underground manifested through a difference of density, might be monitored, providing a new and important database for earthquake prediction.

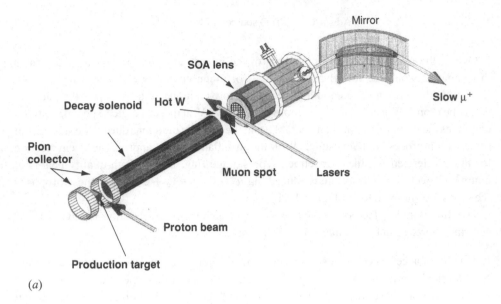

(a)

Mirror

SOA lens

Slow μ⁺

Decay solenoid   Hot W

Pion
collector

Muon spot   Lasers

Proton beam

Production target

Pion capture and
decay solenoid

10keν μ⁺ generator

Laser

Proton
accelerator

μ⁺ Linac

Race track
storage ring

e⁺ dump

(b)

$\overline{\nu}_u, \nu_e$

Figure 10.4 Using intense production of ultraslow μ^+ (a) the scheme of intense neutrino production (b) is proposed based upon a decay of accelerated muons in a storage ring.

10.6 The $\mu^+\mu^-$ colliders with slow μ^+ and μ^-

In contrast to hadron colliders such as $p\bar{p}$ colliders, e^+e^- colliders generate simple single-particle interactions. However, their extension to the TeV region using a circular-accelerator method like a synchrotron is extremely difficult in practice due to an energy loss from bremsstrahlung effects such as synchrotron orbital radiation. Therefore, a linear collider with a pair of full-energy linacs is required. In order to perform collision experiments with lepton pairs in the TeV region to yield a factory of t-quarks or Higgs particles, as an alternative method to the linear e^+e^- collider, the concept of a $\mu^+\mu^-$ collider was proposed in a realistic design (Skrinsky, 1980; Neuffer, 1984).

Muons, with a mass of 207 times electron mass, have negligible bremsstrahlung which depends inversely on the fourth power of the particle mass. Muons can be accelerated in an efficient and smaller circular machine and stored in the ring, leading to $\mu^+\mu^-$ collisions at energies up to the TeV region. Because of the absence of bremsstrahlung, precise measurements of the masses or widths of new particles such as the Higgs boson can be studied with a higher cross-section as compared to e^+e^- colliders. (The cross-section is proportional to mass squared.) The difficulties to be overcome before this type of machine can be realized are as follows: (1) the limited lifetime of 2.2 μs (τ_0) at rest must be overcome by a rapid initial energy increase (to E_μ) ($\tau_\mu = \tau_0 \times E_\mu/m_\mu$) so that the lifetime is 0.044 s at 2 TeV; (2) the decay products (e^+ or e^-) may cause background in the detectors; (3) muons produced by the decay of pions have a large phase space (momentum spread times spatial spread), requiring some drastic cooling method.

In the presently proposed concept of the $\mu^+\mu^-$ collider (Palmer et al., 1996), as shown schematically in Figure 10.5, ionization cooling has been proposed as a realistic cooling method for application to a source of energetic (0.1 ∼ a few GeV) muons which are subsequently to be further accelerated. In this method, isotropic energy degradation through ionization energy loss inside matter under a longitudinal accelerating force is considered to be cooling in the transverse direction.

It is an interesting question whether the ultraslow muon sources described in Chapter 2, in their ultimate technically upgraded forms, can be used as the ion source of a muon collider (Nagamine, 1996); these slow μ^+ and μ^- can contribute to the concept of a $\mu^+\mu^-$ collider, as shown in Figure 10.5. The key factors necessary to judge this possibility are the intensity and emittance. Let us consider the ultimate values of these parameters, and compare them to those for the ionization cooling method proposed for the $\mu^+\mu^-$ collider. Let us take the situation in which the μ^+ are produced in the super-super muon channel with a large acceptance of pions generated by the collision of 3 GeV and 200 μÅ protons with some primary target; these muons are then to be delivered on to multilayers of hot W. From such a set-up, we can expect 10 keV slow μ^+ with an intensity of 10^{11}/s. Since the initial transverse energy of μ^+ is only 0.2 eV, the emittance at 1 TeV is 6×10^{-11} rad/m. These values should be compared to $10^{11}\mu$/s and 10^{-8} rad/m expected to be realized in the ionization cooling method.

As for the production of slow μ^-, by adopting a multilayered $H_2(T_2)$-DT target for the muon catalyzed fusion type of cooling method described in Chapter 2, and coupling this

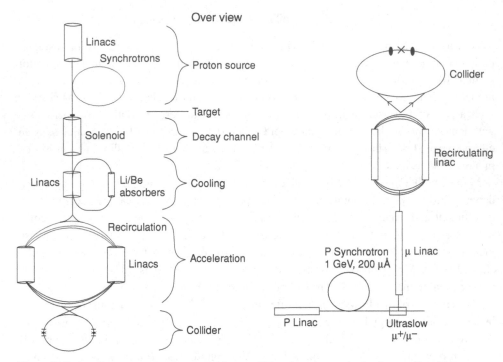

Figure 10.5 (*a*) The proposed scheme of $\mu^+\mu^-$ colliders based upon ionization cooling (Palmer *et al.*, 1996) and (*b*) new scheme for $\mu^+\mu^-$ colliders based upon thermal/zero-energy cooling of the μ^+ and μ^- beam.

with intense MeV μ^- production via the super-super muon channel, we can expect 10^9 slow μ^-/s with an emittance of 10^{-9} rad/m at 1 TeV.

10.7 Mobile TeV muon generator and disaster prevention

As described in Chapters 2 and 9, although the intensity is limited, high-energy (GeV–TeV) muons are produced as secondary cosmic rays by the interaction of the primary cosmic-ray protons with nuclei (N, O, etc.) in the atmospheric air. For the purpose of probing the internal structure of a very large object such as a volcano, the horizontal muons are most suitable for use, provided that the muon flux is reasonably high and that the size of the detection system is realistic. A new prediction method for volcanic eruptions is proposed by using the detection of near-horizontal muons passing through the active part of the volcano, as described in Chapter 9. In order to overcome the intensity-limitation problem, it is indispensable to consider a use of an accelerator. In order to obtain a mobile TeV source, a reasonable way might be to produce muons by some modest low-energy proton accelerator and accelerate only muons produced in this way up to TeV energies employing an acceleration scheme developed for either $\mu^+\mu^-$ colliders or neutrino factory.

Here, we consider the most compact scheme of TeV muon source to be realized by the technologies presently available, and can be mounted on to the "mobile" ship of an

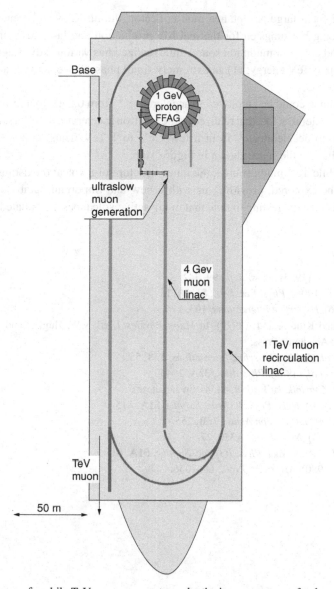

Figure 10.6 Dream of mobile TeV muon source to probe the inner structure of volcanoes: a mobile on-ship TeV muon source. There, 1 GeV fixed-field alternating-gradient of synchrotron (FFAG) proton is used for pion production, followed by the super-super muon channel and the ultraslow μ^+ generator. The muon ion source thus produced is connected to a 4 GeV linac and 1 TeV recirculation linac. All of these components will be on the base of the existing aircraft carrier.

aircraft carrier. The scheme shown in Figure 10.6 is now proposed. As for muon production, a compact proton accelerator such as a fixed-field alternating gradient synchrotron (FFAG) up to 1 GeV could be used (Mori, 1999). There, with a short proton linac, the compact accelerator can be constructed within a circular space of 30 m diameter. Then, following a super-super muon channel for a large acceptance pion collector, the ultraslow μ^+ generation would be installed using hot tungsten for thermal Mu production and laser resonant ionization. The intense and compact muon ion source thus realized has various advantages for further acceleration up to TeV energy; (1) an extremely small phase space; and (2) a small energy spread ($\pm$ 0.2 eV).

A compact linac could be employed to accelerate μ^+ from 0.2 eV to a few GeV. Then, a recirculating accelerator, like that realized as the electron accelerator at Jefferson Laboratory, could be used to accelerate μ^+ from a few GeV to 1 TeV. Thus, hopefully the whole installation will take the space shown in Figure 10.6.

Using a mobile TeV muon source, the inner structure of a volcano existing near the sea can instantly be explored, providing us with a new and important database for eruption prediction. It is also interesting to note that many active volcanoes are situated near the sea.

REFERENCES

De Rujula, A. *et al.* (1983). *Phys. Rep.*, **99**, 342.

Fried, D.G. *et al.* (1998). *Phys. Rev. Lett.*, **81**, 3811.

Hauser, P. (1996). *Hyperfine Interactions,* **103**, 175.

Hughes, V.W. and Kinoshita, T. (1977). In *Muon Physics 1*, ed. V.W. Hughes and C.S.Wu p. 11. New York: Academic Press.

Kottmann, F. *et al.* (2001). *Hyperfine Interactions,* **138**, 55.

Ma, Qian-ching *et al.* (1985). *Phys. Rev.*, **32A**, 2645.

Mori, Y. (1999). *Genshikaku-kenkyu*, **44**, 41 (in Japanese).

Nagamine, K. (1996). *Nucl. Phys. B (Proc. Suppl.)*, **51A**, 115.

Nagamine, K. (1999). *Proc. Jpn Acad.*, **75B**, 255.

Neuffer, O.V. (1994). *Nucl. Instr.,* **A350**, 27.

Palmer, R. *et al.* (1996). *Nucl. Phys. B (Proc. Suppl.)*, **51A**, 61.

Skrinsky, A.N. (1980). *AIP Conf. Proc.,* **68**, 1056.

Further reading

Hughes, V. W. and Wu, C. S. (eds) (1977). *Muon Physics*, vols. 1–3. New York: Academic Press.

Karlsson, E. B. (1995). *Solid State Phenomena: As Seen by Muons, Protons and Excited Nuclei.* Oxford: Oxford University Press.

Lee, L. S., Kilcoyre, S. H., and Cywinski, R. (1999). *Muon Science: Muons in Physics, Chemistry and Materials.* Bristol: Scottish University Summer School in Physics and Institute of Physics Publishing.

Schatz, G. and Weidinger, A. (1996). *Nuclear Condensed Matter Physics: Nuclear Methods and Applications.* Chichester: John Wiley.

Schenck, A. (1985). *Muon Spin Rotation Spectroscopy.* Bristol: Adam Hilger.

Walker, D. C. (1983). *Muon and Muonium Chemistry.* Cambridge: Cambridge University Press.

Index